OXFORD RESEARCH STUDIES IN GEOGRAPHY

Chemical Foundations

The Alkali Industry in Britain to 1926

Kenneth Warren

CLARENDON PRESS • OXFORD
1980

Oxford University Press, Walton Street, Oxford OX2 6DP

OXFORD LONDON GLASGOW

NEW YORK TORONTO MELBOURNE WELLINGTON

KUALA LUMPUR SINGAPORE JAKARTA HONG KONG TOKYO

DELHI BOMBAY CALCUTTA MADRAS KARACHI

NAIROBI DAR ES SALAAM CAPE TOWN

Published in the United States by
Oxford University Press, New York

British Library Cataloguing in Publication Data

Warren, Kenneth, *b. 1931*
Chemical foundations. – (Oxford research studies in geography).
1. Alkali industry and trade – Great Britain – History
I. Title II. Series
338.4'7'66130941 HD9660.A43G7 79-41050

ISBN 0-19-823231-4

Set by Malvern Typesetting Services Ltd
and printed in Great Britain by
The Fakenham Press Ltd, Fakenham

PREFACE

Disraeli was once derided for quoting renewed activity in the chemical industries as a sign that the trade cycle was on the upturn. At the time the chemical trades covered a very much narrower field than now and were of relatively small importance within the economies of the advanced industrial nations. Yet Britain then led the world much more decisively in chemicals than in most other industrial activities. She is now far outstripped, producing in the mid-1970s, for instance, about 4 per cent of world sulphuric acid. At the beginning of the final quarter of the twentieth century even the heavy chemical industry comprises a bewilderingly wide range covering sulphuric acid, ethylene and its associated products, plastics and fibres and ammonia and other fertilizer chemicals. A century ago, the bulk chemical industry was essentially the alkali industry. Diversification of products, new sources for old raw materials, and, more important, wholly new production techniques and materials have changed the structure and location of the industry. Because of these changes there has been less continuity than in, for instance, steel, but old patterns of location can still be traced amidst the present structure of the industry. In the 1970s the bulk chemical industry in Britain uses only two home raw materials in large tonnages: salt and limestone. Petroleum is the largest single input, and imports of sulphur or sulphur-containing minerals are also important. Coal, pre-eminent not only in the nineteenth century but even up to about 1960, is now of very minor significance. Yet, despite all these changes, the Mersey basin contains in Widnes Britain's leading single centre of sulphuric acid production; Widnes and Runcorn have by far the largest share of British chlorine capacity; and the whole of Britain's production of soda ash and much of its production of caustic soda is in the Cheshire salt-field or the middle Mersey district. Eighty years ago this was the world's prime focus for the alkali industry. Teesside holds the lead in Britain's ammonia and associated fertilizer capacity and, to a lesser extent, in the production of olefines. In this area, too, there is a link, albeit much more tenuous, with the old alkali industry. As well as shaping the modern map of chemicals, the alkali industry had a vital role in the making of the landscape in various parts of Britain, not only in present chemical districts, but also in localities from which the industry has disappeared, such as St. Helens, parts of east Glasgow, and, above all, Tyneside. Its effects were wider still. The Buxton limestone workings, now so troublesome to those concerned to preserve the amenities of the Peak District, were opened largely to supply Lancashire's alkali trade, and Sicily's wasted sulphur districts or such famous Iberian mining centres as Rio

Tinto or Tharsis were also tributary to British alkali manufacture. As the world-famous mining concern named after Rio Tinto shows, the effects rippled outwards into quite different industries as well as into remote areas.

The chemical industry has been neglected by most economic historians, though in recent years there have been some noteworthy new developments. It has been largely ignored by geographers, even though it shaped regions and landscapes and exemplified the various characteristics and factors affecting locational change quite as well as iron or cotton or ship-building. It is in the belief that this deficiency should be made good that the following study of a small part of this vast industry over part of its history has been made. The field for historical study of industrial locational change is large; there is room for many cultivators.

I would like to acknowledge the freedom with which I.C.I. gave me access to the records of their alkali-making predecessors, both in London and in Widnes. They have also been generous in providing illustrations.

Witney, Oxfordshire KENNETH WARREN
March 1979

CONTENTS

LIST OF ILLUSTRATIONS

FIGURES

PLATES

(between pp. 42–3)

ABBREVIATIONS

The titles of journals etc. are abbreviated as follows:

British Parliamentary Papers	*B.P.P.*
Chemical News	*C.N.*
Chemical Trade Journal	*C.T.J.*
Colliery Guardian	*C.G.*
Engineering	*Eng.*
Iron and Coal Trades Review	*I.C.T.R.*
Journal of the Iron & Steel Institute	*J.I.S.I.*
Journal of the Society of Arts	*J.S.A.*
Journal of the Society of Chemical Industry	*J.S.C.I.*
Journal of the Statistical Society (later the Royal Statistical Society)	*J.S.S.*
Mining Journal	*M.J.*
Transactions Newcastle-upon-Tyne Chemical Society	*T.N.C.S.*

INTRODUCTION

Since World War II about a dozen major oil-refining centres have been established around the coast of Britain, some in wholly new locations, others at small inter-war refining or oil storage points. From the late 1950s some developed into petrochemical centres. They are now the biggest agglomerations of modern British chemical capacity in terms of volume of output, employment, or capital investment. There is, however, another, older pattern discernible in the complex chemical map of Britain. Sometimes this reflects widespread demand and localized materials as in the fertilizer industry and the part-associated sulphuric acid industry. Sometimes it is connected with what is virtually a by-product, as in the case of chemicals associated with the metal trades or with coke ovens, as at integrated steelworks or a handful of coal-mines. For the rest, the heavier side of the British chemical industry still reflects Merseyside's pre-eminence in the alkali section of the heavy inorganic side of the business. In a wider sense, despite the very great technological, locational, and organizational changes of the last few decades, the alkali industry can be regarded as the germ from which the whole modern chemical industry has grown.

'During the first half of the nineteenth century, and even later, the alkali industry was the chemical industry,' stated two recent historians of the industry (Hardie & Pratt, 1966, p.27). As early as the 1880s, however, practitioners recognized that their industry was already widening: 'Chemical manufacture has now ceased to have the limited meaning that was formerly attached to it—it is no longer "the alkali trade" with a few etceteras.' (Proctor, 1883, p.298.)

Not only did alkali manufacture decrease in relative importance, but within it technical change brought the demise of one process and the emergence of a rival one. It was not, however, simply a matter of technical change, it was also a response to opportunities and challenges. As Ernest Solvay remarked in 1885, at the Silver Jubilee of his own highly successful firm, 'Industries like men grow old, and in celebrating their silver wedding you are too often only celebrating their decline' (Reader, 1970, p.95).

Even in its decline the older processes helped to spawn a range of new chemical activities. Forty years after Proctor, in what was effectively a valedictory on the old industry, this process of adjustment, of innovation, of branching out into new trades was recognized—'the Leblanc process is completely discarded, but the industry which was built upon it remains, still advancing, developing in ever new directions, demanding still of those concerned in it the same qualities, judgment, audacity and untiring labour . . .' (*U.A.C.*, 1923, p.46). In another sense too the alkali trade is of vital significance for an understanding of the modern British chemical industry. Imperial Chemical Industries, created in 1926, was recognized as Britain's best hope of meeting the competition of newer chemical industries overseas, particularly in Germany and the United States. Two of its four constituent companies were the leaders in the struggle between old and new alkali techniques—the United Alkali Company and Brunner Mond & Co. In short, two of the four bases on which the contemporary British chemical industry was built were alkali firms.

In one other aspect the old alkali industry has shaped the British landscape—through its effect on the character of industrial plants and the associated urban areas. In 1928 a commercial journalist described the large, new, and efficient chemical works and urban foci which were to emerge in Britain in the middle decades of the twentieth century. The model had long existed, in Ludwigshafen or Leverkusen, or even in Syracuse, or to some extent at Northwich. 'We dare to suggest that most of the great chemical works of the future will be of the type that Billingham is becoming. Instead of a hundred or so sulphuric acid works dotted all over the country, some in the midst of congested areas with no room to expand, we shall yet see a coalescence into large ordered units. . . . the future "chemical town" as we see it will be a town on its own, containing from 5,000 to 10,000 workmen, or, say, 15,000 to 30,000 inhabitants, together with a housing estate with shops or co-operative stores in some form or another.' ('Dimethyl Town, England', *Industrial Chemist*, 1928, p.1.) A few years earlier the great contrast with the old pattern of chemical trade and chemical town development had been provided in the description of what Partington (1919, p.2) labelled 'alkali towns'—'the rows of chimneys emitting black smoke from the unscientific combustion of coal, the enormous lead chamber towers, revolving furnaces, waste heaps with escaping steam, the noise, and the smell of acids, chlorine and sulphuretted hydrogen, are all familiar'. Partington might have added something on the drabness of the urban fabric itself—the black terraced houses, the solid but dull Board Schools, Gothic revival churches, and variously styled but generally architecturally poor civic buildings, and poverty of public service provision. More than half a century later there is a new, familiar chemical scene—at Billingham, Immingham, and Avonmouth, from Grangemouth to Baglan Bay and Fawley—landscapes of columns, of convoluted pipework, of massive, open, sprawling complexes, not without their problems of pollution but certainly not black and gas- and smoke-drenched. The good semi-detached and detached housing of Billingham has given way to longer journeys to work, so that post-war Wilton, for instance, has no nearby closely-knit chemical community, but, whether concentrated or scattered, there is a remarkable contrast with Victorian chemical towns. Unfortunately elements of the old pattern are still very much in evidence, even though the chemical trade itself has sometimes wholly disappeared—as in parts of East Gateshead or Hebburn, in Oldbury, in Runcorn in spite of great town expansion and reconstruction, and also, and pre-eminently, in Widnes. Time and again in modern Britain the discerning traveller will find elements of the old alkali industry.

The national significance of the alkali industry

Though the nineteenth-century alkali industry has been of considerable importance in shaping the economic geography and still more the cultural landscape of modern industrial Britain, throughout the Victorian age it was overshadowed by other trades. In the late 1830s, McCulloch (1839, p.732) summed up his survey of the industries of Britain: 'Such is a short sketch of the more important statistical details connected with the principal manufactures carried on in the country', but he made no mention of alkali. Over forty years later when Leblanc alkali was at its peak level of production, the publishing firm of Cassells (1882) produced a three-volume survey of the basic industries of the United Kingdom. It contained very little on chemicals, and only one plant, St. Rollox in Glasgow was singled out for description. There were, however, snowball effects derived from chemical manufacture, through the alkali-using trades, through the suppliers of raw materials and services, and in the satisfying of the consumer demand of the considerable work force. Though unable to analyse it in any detail, some Victorians were aware of this multiplier effect. Quoting Liebig's statement that the Leblanc process was the foundation of all modern improvements in the domestic arts (an excessive claim), Huskisson (1853, p.441), lecturing on salt, sensed this wider importance: 'It is of immense importance as a source of soda. Soda has been used from time immemorial in the manufacture of soap and glass—two chemical productions which employ and keep in circulation a vast amount of capital.' Ten years later, in the House of Commons, Vivian noted that the alkali trade annually consumed 1.76 million tons of raw materials, and produced 0.28 million tons of finished goods worth £2.5 million. Wages paid were £0.87 million (*M.J.*, 18 July 1863). The industry's capital was about £1,000,000 in 1852, £2,000,000 ten years later, and by the mid-1880s £10,000,000 (ibid. and Allhusen, 1886, pp.264–5). The workforce was reckoned to be 19,000 in 1863; by 1880 it was 40,000. 'Few [industries] have had a more rapid rise, and fewer perhaps have attained to greater importance,' remarked a leader writer in one of the trade journals when looking back on this progress from the recession which had followed it (*Industries*, 29 July 1887, pp. 119–20). With the industry rapidly declining, Gibson (1893, p.91) remarked on the superstructure built upon it, 'the magnitude of the capital invested, of the volume of the yearly turnover, and of the army of labourers employed, cannot fail to impress; nevertheless, these aggregates are relatively unimportant in comparison with the vastness of the other industries whose prosperity depends on the cheap supply of the products of common salt.' Yet even when this wider scope is taken into account, it remains true that the Victorian chemical industry, world-dominant though it was, was a stripling among industrial giants. The Royal Commission on Coal in 1871 showed that of every 1,000 tons of coal raised in Britain, 300 went to iron- and steelworks, 60 to gasworks, 42 to textile works, and only 19 to chemical manufacture (Bell, 1873, p.17).

Alkali: an industry, eotechnic, palaeotechnic, and neotechnic

In the early 1930s Mumford (1934, pp. 109–10), elaborating earlier work by Patrick Geddes, stated that 'looking back over the last thousand years, one can divide the development of the machine and the machine civilisation into three successive but *overlapping* and *interpenetrating* phases: eotechnic, paleotechnic and neotechnic'. In power and characteristic materials, 'the eotechnic phase is a water and wood complex; the paleotechnic phase is a coal and iron complex; and the neotechnic phase is an electricity and alloy complex'. With each went a 'larger social pattern'. Different raw materials and techniques imply also different types and scales of organization and production and changing locational patterns. Mumford saw this pattern of change as extending over centuries: in heavy chemicals it can usefully be applied to the changes of the nineteenth century. At the beginning, alkali was derived from vegetable matter, by labour-intensive gathering and widely scattered burning, on Mediterranean coasts, in northern woods, or on the shores of north-west Britain. At the end of the nineteenth century came the neotechnic operations of the ammonia soda process, large-scale, with elegant processes, producing little waste, and dominated by a few big organizations. Even more strikingly neotechnic was the electrolytic manufacture of a limited range of alkali products, notably caustic soda and bleaching powder, using brine, electricity, and limestone as the main inputs. Between the gathering and simple processing which dominated the first quarter of the century, and the Solvay and electrolytic processes which together were the technical and commercial triumphs of its last quarter, came a half-century dominated by the Leblanc process. This was an almost archetypal palaeotechnic industry, involving large tonnages of raw materials, a great loss of weight in manufacture, and, as the corollary, production of immense amounts of waste—gaseous, liquid, and solid. The essential steps in the process remained more or less the same throughout the century, but in the early days there was so much inefficiency in the system of production that extremely large economies were made over time. Partly for this reason, the Leblanc process was able to adjust to the competition of the neotechnic processes by improvements in efficiency which frequently involved large-scale use of what had formerly been waste products, and which eventually led to a transference of the attention of the trade to these to the neglect of its earlier alkali specializations. With evolving techniques went changes in organization, the elimination of many small firms, the pre-eminence of the bigger, and, eventually, a combination of almost the whole trade which produced what could briefly be claimed to be the world's largest business organization. Business methods, attitudes to research, relations between masters and men—all changed, from what seemed the epitome of the classical capitalist enterprise to something which was suitable as a host on which to graft a twentieth-century growth stock.

The business and technical uncertainties within which the palaeotechnic alkali industry operated can nowhere have been better exemplified than in the friendly letter written on behalf of the Netham Chemical Company, Bristol, to the Midland firm, Chances of Oldbury, in November 1876, concerning caustic soda manufacture (Worsley, 1876).

> Dear Mr. Chance,
>
> I hope you are not going into such a valley of sorrow as caustic making has been to us. I can only say that if you can do without making so capricious an article you had better not touch it. As to figures, like yourself I have nearly ceased attaching much weight to any I am told; and as I have no right to suppose other people are any less veracious than the Netham Co. I can only say, do not attach more value to ours than they deserve. Answer your questions to your satisfaction I cannot—because as regards production, we do not know it. It is purely a matter of speculation, and we keep no account of it. Lime and nitre are always varying according to circumstances from time to time. We never get two half years alike. Fuel also varies. In truth caustic making is a matter we have to be hopeful about and bear it as we can. It takes a heap of steam, which we cannot measure and is a tangle of unknown quantities. We are just now wanting to find a way of dealing with the caustic sludge and the caustic fished salts—both very awkward customers. Are you open to buy them—the latter. They contain about 30% caustic, the residue being sulphate and chloride, mainly cloride and a heavy proportion of iron.
>
> I suppose that if we were to go regularly into caustic instead of making it only as a sort of by-product, we should get more uniformity in our results. I believe the proper way is to make caustic from vat liquor, whereas we have hitherto used our red liquors with all their drawbacks and variations and hence our difficulties. I am sure people who make caustic only must do better than we do.
>
> Yours very truly,
> PHILIP J. WORSLEY

Along with rule-of-thumb manufacturing went an old exploitative attitude to labour, though there were some exceptions. Chance for many years maintained 'a convalescent home on a breezy hillside, where the sick ones may regain their strength'. Later he provided a recreational ground near the works. In St. Helens Kurtz put up baths and washhouses and Gamble helped liberally in the provision of schools (*B.P.P.*, 1893–4, xvi). There was strong distaste for the dull routine of the chemical laboratory and a firm adherence to practical rather than theoretical insights into the business, though operators such as Henry Deacon in Lancashire or Alexander Chance in the Black Country made impressive contributions to the chemistry of alkali manufacture. A. E. Fletcher, government inspector under the Alkali Acts, recalled his reception at one large north-western chemical works in 1864: he was taken into the chemical laboratory and introduced to the chemists and managers by the general manager as 'the gentleman who had come up from London to teach them their business' (*C.T.J.*, 5 May 1888). In the mid-1870s, in an extensive examination of the alkali trade, Kingzett (1877, pp. v, 6, and 8) noted that second-rate chemists were frequently employed. He too had a rather derogatory view of the role of the laboratory, 'accidental discoveries are somewhat inevitable in experimental science, where the method of enquiry depends upon reasoning, for the correctness of which there is no absolute test'. Since most manufacturers were not scientific men, and improvement involved writing off capital, 'the introduction of recent improvements in the alkali trade was attended with much opposition from manufacturers themselves'. At about the same time John Morrison, spelt out the lack of science and of professionalism in an address to the Newcastle upon Tyne Chemical Society (Morrison, 1875a, p.87).

> I have frequently lamented the apparent absence of identity about the status of a chemical works manager. It seems quite a question whether our calling ought to rank as a profession, or whether we do not rather belong, plainly and simply, to a superior order of works detective. We are a class of men called into existence, as most other classes of men, by Mother Necessity, but occupying a somewhat abnormal position, (I do not speak of the Tyne especially); and probably from the peculiarity and the variable character of our employment, there has not yet been devised any special course of training for aspirants to our office. . . . Many of us have gradually taught ourselves to look upon the laboratory almost in the light of a necessary nuisance, an institution only of use to us to ascertain if we get full value in our raw materials, and that our finished products are sent out up to strength. Henceforward, we have said to ourselves, our chemistry must be confined to productions and costs per ton, and, in Lancashire especially, to quantity turned out per furnace; the best manager being he who, prima facie, can get the greatest number of tons out of a given number of men and a given amount of plant.

Chemical science was nevertheless making progress, and less than a decade later, in a review of past years on the Tyne, Proctor (1883) recognized that 'chemical law now reigns in many works which were formerly ruled by thumb'.

Slowness in recognizing the science as opposed to the mere brute force of chemical production distinguished the British alkali industry from that on the Continent well before Britain lost its lead in production. The superiority of Britain in output had attracted Continental chemists, so that in the seventies there were 'shoals of poor German scientists who infest this country' (*M.J.*, 28 Apr. 1877, p.465). Among the immigrants from the Continent were Ludwig Mond, who was appalled by the wastefulness of the Leblanc firms. Another was Ferdinand Hurter, who made impressive discoveries and became one of the staunchest defenders of the Leblanc process against the attacks which Mond's genius spearheaded. Georg Lunge, after studying under Bunsen at Heidelberg, and then at Breslau, came to the Tyne in the 1860s. When he applied for a post to the senior partner of one of the largest works on the river, he was offered a job at £1 per week, or only 2*s*. more than the labourers (Clapham, 1868, p.354). (It is, however, only fair to add that at the same time Henry Roscoe, professor of chemistry at Owens College, Manchester, pointed out that chemical works in the North-West were open to educational visits, and that in those works, chemists were widely employed and well paid (Roscoe, 1868, p.278).) Generally efficiency at British works was then visibly falling behind that on the Continent. There was a temptation for management to blame the workmen for this. In the early 1870s, the Tyneside manufacturer R. C. Clapham visited some of the French Leblanc works. He found the raw materials more costly, and wages and productivity both less than in England. The product was of better quality. Men in the sulphuric acid departments were paid 3*s*. a day as compared with 5*s*. in England. Yet in twenty-four hours the saltcake furnaces worked only $2\frac{1}{2}$ tons of salt as opposed to 10 tons in English works. The workmen were 'industrious, frugal and orderly, drinking being little known, and they were more subject to discipline and the control of their foreman than with us'. In England it seemed to him that an emphasis on quantity of production was in part a response to high labour costs, whereas across the Channel, 'the constant application of the chemical laboratory to every process, and care of the workmen are producing dearer but superior products to the English' (Clapham, 1874, pp.275 and 276). Three years later, Kingzett, though noting that British manufacturers were much more 'practical' than 'scientific', stressed again the importance of volume of production and highlighted the powerful advance of the German industry even before the new Tariff Act of 1879 gave it the necessary protection for a larger increase in output. One of the largest chemical works in Germany already employed six chemists at from £300 to £400 per year and 'in addition retains the services of a chemist of reputation, *exclusively for theoretical work* in the laboratory, at a salary of nearly £2000 per annum' (Kingzett, 1877, pp.10, 234, 235).

In the crisis that faced the Leblanc industry in the eighties, it became clear that there was a need for its firms 'to stir up those faculties of invention and enterprise which they have allowed to remain dormant so long' (*C.T.J.*, 28 May 1887, p.3). 'In France and Germany we are bound to confess that all chemical operations are carried on more carefully than they are in England, and this is the case in America too; more attention is there paid to detail, whereas in England we go on in the same clumsy manner as did our forefathers fifty years ago.' (*C.T.J.*, 4 June 1887, p.10.) 'What the English alkali makers will have to learn is that they must throw over their old rule of thumb working and place themselves in the hands of good chemical engineers; the work must be done scientifically, and not left so much as it is even now, in many cases to the prejudices of Tom, Dick or Harry.' (*C.T.J.*, 9 July 1887, pp.50–1.) Improvements were made, and in 1890 the United Alkali Company began a rationalization process, but the lead of the Continental producers could not be removed. By the mid-1890s the president of the Society of Chemical Industry described the impact made by German chemical excellence: 'One of the greatest object lessons in so-called technical education that I know of is to be seen on the banks of the Rhine at Ludwigshafen, and I recall as if it happened yesterday, the vivid impression made upon me by the sight of the wonderful works of the Badische Analin und Soda Fabrik.' (Thorpe, 1895, p.156.)

Alkali and the spirit of the age

The alkali industry made fortunes for its capitalists and a drab, sometimes steady, sometimes unhappy and uncertain, employment for its workforce, many of whose members died of industrial diseases. To the Victorians it all seemed part of the nature of things, the reward and price for progress; wealth, optimism, sometimes vision, and a driving materialism were the order of the day. Manufacturers made their money and often spent it outside the region. When John Williamson of the Jarrow Chemical Works, South Shields, died in 1887 his personal estate was only £45,000, but his address was the Villa Guiseppona at Cadenabbia on Lake Como (*C.T.J.*, 5 Nov. 1887, p.212). James Muspratt, pioneer of large-scale working of the Leblanc process, retired in wealth to live out a long life in his house at Seaforth, well away from the smell of his works at Widnes, Flint, or along the Vauxhall Road. Christian Allhusen of Gateshead played a part in local commerce, government, and society; he bought Elswick Hall from Richard Grainger, a renowned Newcastle speculative builder, but in retirement occupied Stoke Court near Stoke Poges, Buckinghamshire, a not untypical example of the transfer of wealth from industrial regions to the softer south. At his death in 1890 his personal estate was £1.1 million. John Tennant, son of the founder of the great Scottish alkali firm, had £1 million when he died in 1878. His son Sir Charles Tennant left a fortune of over £3 million in 1906 (U.A.C., 1923, pp.57–9). Hugh Lee Pattinson of the Felling Chemical Works 'in his latter years liberally devoted to the cultivation of astronomy much of his leisure and of the fortune acquired in the alkali works he established on the Tyne' (Gibson, 1860, p.7). Pattinson died in 1858. Five years earlier he delivered the address at the first soirée of the Felling Chemical Works Reading Society. It epitomized the confidence of unchallenged industrial leadership.

> It is impossible to contemplate the progress of science within the present century, without feeling that a new epoch has commenced in the history of our race. The powers of the human mind are extending their grasp. Fields of knowledge, undreamt of in the earlier ages of the world, are successfully cultivated. The hidden forces of nature; the laws by which her phenomena are governed in their endless variety and succession; the structure and properties of matter; all are undergoing a vigorous process of scientific investigation, from which result discoveries that would be deemed miraculous did not their number and frequency almost exhaust our faculty of wonder. The great characteristic of this grand intellectual movement is its practical tendency. Learning and skill are esteemed as they conduce to the well-being of society, and their value is measured by the extent of their application to the practical affairs of life. (Pattinson, 1853.)

Such a philosophy of material progress may have seemed satisfactory to Pattinson in the comfort of his house at West Boldon away from the Tyneside industries, but it can have been of little help to those who lived in Felling Shore or Heworth or Bill Quay, in miserable homes pinched in between works or expanding waste heaps and a river which was heavily polluted and often shrouded in smoke or chemical fumes.

Later, as competition both at home and overseas became keener, the principles of Social Darwinism became popular. New techniques and producers swept old methods and districts out of existence. In 1886 the Chairman of the Glasgow Section of the Society of Chemical Industry observed that 'he had no doubt that the depression which we had all been experiencing for some time past will have good effects. It will lower costs all round, and be the means of enabling them to compete with other countries, and lead to what always helped in that and all other industries, the survival of the fittest.' (*J.S.C.I.*, 29 Nov. 1886, p.590.) Others found in the crisis a materialist rationale for more education. In autumn 1888 E. Muspratt spoke at a prize-giving at Widnes Science and Art School: 'The time had come (unless the trade of Widnes was to depart from it, and the great manufacture of alkali, copper extraction, and the other trades carried on in Widnes were to be transported to other countries), when they must give a higher and better education.' (*C.T.J.*, 24 Nov. 1888, p.325.) Walter Weldon was a complex character, an amateur chemist, a journalist, and a man in whom, as Alexander Chance put it, 'science never operated to chill the heart', but even he was persuaded to see in a relentless if painful material progress some promise of later, greater good. In 1884, when recognizing that as a means of producing soda the Leblanc process had served its time, Weldon concluded, 'By such changes, individuals may, and alas do, suffer but the world gains by them' (*J.S.C.I.*, 29 July 1884, p.387). His obituary highlighted one of the outcomes of the Victorian preoccupation with material progress. 'Like so many in this strenuous generation, Mr. Weldon suffered from overwork. Never satisfied with what he had done, he entertained larger designs, labouring unremittingly for their development. Continually warned that the strain on his energies was excessive, he held on in weakness as in strength.' (*J.S.C.I.*, 29 Oct. 1885, p.580).) Weldon was only 53. John Glover, though he lived to the age of 85, was also subjected to the rush of Victorian industrial life. The Glover tower which denitrated and concentrated sulphuric acid was one of the vital inventions of the alkali trade. Glover visualized it in 1847, but was too busy to build the first tower until another twelve years had passed. (*J.S.C.I.*, 15 May 1902, p.595; *Endeavour*, I, 1942, pp.42–3.) In the 1890s it was reckoned to save the industry £0.3 million annually. As Reader (1970, p.122) has remarked, the Leblanc alkali industry 'in many ways seems the supreme example of that decadence in late Victorian British industry which alarmed some contemporaries, and has since fascinated economic historians'.

CHAPTER 1

The Demand for Alkali

The demand for alkali which gave rise to its large-scale production came both from a wide range of not very closely associated manufacturing industries and from direct consumption by the population at large. Some products, such as soda crystals (washing soda) or bicarbonate of soda, had a consumer demand whose distribution reflected that of the population. The uneven distribution of such major consuming industries as the glass, soap, paper, and textile trades affected the marketing of other alkali products. As Britain dominated the world market for alkali until well into the last quarter of the nineteenth century, export outlets were also of vital significance.

Soap

Soap was used increasingly in both manufacturing processes and personal hygiene as industrialization proceeded. Liebig's 'Familiar Letters on Chemistry' argued that soap consumption provided an index of progress: 'The quantity of soap consumed by a nation would be no inaccurate measure whereby to estimate its wealth and civilization.' (*Penny Magazine*, 28 Oct. 1843, p.416.) Others more realistically made allowance for human nature. A Frenchman writing about the Marseilles soap industry adopted a tone of disillusionment, 'All this is made for the purpose of keeping people clean, yet despite the advance of science they remain dirty' (*C.T.J.*, 28 Nov. 1891, p.334). Though hindered by an excise duty of varying level from 1712 to 1853, the British soap trade increased rapidly. Soap-making had been a widely spread and predominantly small-scale activity. Entrance to the trade was easy, and exit frequent. In 1833 there were 262 licensed soap-makers in England. Over the next fifteen years another 344 new firms took out licences, but of the combined total of 606 firms only 157 were still in business in 1848 (Hawes, 1856). Figures provide only a rough indication of the progress of the trade, for a great many people were known to make soap without a licence (Porter, 1836). Soap was of two main kinds: soft for industrial processes, and hard mainly for personal use. The latter was initially largely made from potash lyes, though some soda was also added. London was the main centre of soap-making before 1800, but the growth of soap-boiling on Merseyside during the eighteenth century soon challenged this pre-eminence. Merseyside soaperies benefited from easy access to coal, good facilities for the importation of crude alkali in the form of kelp and barilla, and extensive trading connections which provided supplies of tallow and palm oil. By the decade 1810–20 kelp was the main source of alkali for the Merseyside soaperies, though barilla, richer in soda and requiring less coal in its use, was naturally preferred in London. In 1822, despite protests from Merseyside, the import duties on barilla were reduced. But the progress of the North-West could not be held back, and Merseyside in the early 1820s made twice as much soft soap as London, and by 1835 50 per cent more hard soap. (Musson, 1965, pp.13, 14, 41, 42.) Demand from the soap-works was an important stimulus to a Lancastrian alkali industry. A number of soap-boilers became alkali-makers, and later alkali firms sometimes went into the soap trade.

Although Merseyside and London led, and the other 'principal seats' of English soap manufacture by 1840 were as in Table 1, there were also 'manufactories of considerable extent at Bromsgrove, Newcastle, Gateshead, Warrington and Plymouth'. In Scotland two-thirds of the soap made was from Glasgow or Leith. (*Penny Cyclopaedia*, 2, 1842, article 'Soap'.)

Table 1. *Soap production in the 'principal seats' in England 1840* (thousand tons)

Liverpool	46,104
London	37,548
Runcorn	12,340
Brentford	5,071
Bristol	7,814
Hull	4,825

Source: *Penny Cyclopaedia*, 1842.

Glass

Alkali and silica sand were the two main raw materials of the glass trade and this was a significant factor in the early localization of the alkali trade (Daumas, 1969). Glassworks needed alkali relatively free from iron. Generally at first, they used soda ash, but in the 1830s and 1840s many turned over to the use of sodium sulphate (saltcake) which was cheaper (Barker, 1960, p.130). As with soap, there was a good deal of common entrepreneurship, though, by contrast, demand grew much more slowly. In the early modern period, when it used charcoal, the woodlands of the Surrey–Sussex border and of Kent were important in influencing the location of English glass-making. In the 1620s a major relocation began when Sir Robert Mansell, who had made glass in Southwark using Scottish coal, moved to Tyneside. Twenty years later other Thames-side glass-makers removed to the Ouseburn industrial district on the east side of Newcastle, and later that century new glassworks were built at South Shields and at Howdon Pans. By the early nineteenth century Tyneside stood out as the main centre for the bulk manufacture of glass—'Newcastle and South Shields are unquestionably the principal seats of the manufacture of crown and bottle glass; the duty paid on the glass manufacture in 1832 being nearly equal to half the whole duty collected in England.' (Leifchild, 1843, *passim*.)

Tyneside's main advantage for glass manufacture derived from its cheap coal, notably the 'smalls' left after the better coals had been exported. On the other hand there seems to have been a notable lack of quality or of entrepreneurship in its glass trade. Its bottle glass was very poor, the ingredients sometimes being merely lime and sea sand wetted with sea-water to provide the necessary alkali. As Henry Chance, the noted Midland glass-maker described it, the process was simple: 'The black bottles of Newcastle are made from common rock salt, and sand from the bed of the river, with carbonate of lime of the soap works, and the tank waste of the alkali makers, but for all better kinds of glass, the circuitous combination of silica and alkali is still found necessary.' (Chance, 1856, pp.222–31.)

Another Tyneside staple was crown glass for windows. Eventually sheet glass was introduced on the Continent as a rival form of window glass, and was indeed long known in the British market as German sheet glass. It was superior in that it could be made in much larger sizes, was free from the waves and lines of crown glass, and did not have the latter's so-called 'bull's eye'. Sheet glass making was introduced to Britain in 1832 by Chance in the West Midlands, and a little later was taken up by Hartleys of Sunderland. It was soon to contribute to changing the rank order of glass-producing centres. In 1833 there were 126 glassworks in the United Kingdom:

> The principal seat of the manufacture in England is at Newcastle-upon-Tyne and the neighbouring town of Shields; next in importance stands Stourbridge; then the works in and near Liverpool, including the Plate-Glass Company's works at Ravenhead; next follow Bristol, Warrington, Birmingham and Leeds; in London there were only three glass houses yielding to the revenue about 2% of the whole amount of duty collected upon this material. In Scotland five out of the ten houses are in and near Glasgow, two are in Leith, the remaining three are at Cartsdike (Greenock), Portobello and Alloa. In Ireland four manufacturers are in Dublin, two each in Cork and Belfast, and one each in Waterford and Newry. (*Penny Cyclopaedia*, XI, 1838.)

In 1845 the glass duty was abolished. Sir Robert Peel, proposing its abolition, anticipated a great future, 'if you leave it altogether disburdened, as in Bohemia, France and Belgium—if you give full and unrestricted freedom to the capital and enterprise of this country—with its peculiar advantage of material—the command of alkali and coal—my belief is you will supply almost the whole world' (quoted Howard, 1866). Repeal did indeed stimulate growth, but, linked as it was with technical change, above all with the triumph of sheet glass over crown glass, and with rationalization in the structure of the industry, it contributed to a major locational shift within the nation. After the repeal, glass houses, as Henry Chance put it, 'sprang up like mushrooms', the output of sheet and crown glass increasing threefold by the mid-fifties, but thereafter the number of producers fell away rapidly. There were 14 firms in the trade in 1844, 24 by 1847, but only 10 in 1854. (Chance, 1856, p.227.) Tyneside as the centre of the older techniques suffered most from this reorganization. In spite of its cheap coal, it was ill placed to supply western British markets with such a fragile product, and when glass prices fell in the depression which followed the building boom of the 1840s, its crown glass works proved no longer able to compete. There was a great crisis in 1849 and 1850. By 1863 Tyneside made no crown glass of any kind. (Barker, 1960, pp.119–21.) Within the country as a whole, by the late 1870s crown glass production could be described as 'practically a thing of the past' (*Encyclopaedia Britannica*, 1879). The Tyneside bottle trade also shrank, so that, as with soap, the local alkali firms lost a major outlet. Again it was Merseyside which became the major growth area.

Paper

Whereas glass manufacture used soda ash or alternatively, and at the end of the nineteenth century largely, sulphate of soda (saltcake), paper production provided an important outlet for caustic soda. Paper manufacture from rags, from straw, or from esparto grass, and later from wood pulp, was fairly widely scattered through Britain though there was a marked process of concentration into fewer bigger production units (Table 2).

Table 2. *United Kingdom paper-mills, 1838, 1850, 1860, 1907*

	1838	*1850*	*1860*	*1907*
England and Wales	*416*	*328*	*306*	*207*
Scotland	*49*	*48*	*52*	*59*
Ireland	*60*	*39*	*26*	*7*
Total	525	415	384	273

Sources: *J.S.S.*, 24, 1861, p.287; *Encyclopaedia Britannica*, 1910/11.

Throughout the nineteenth century much of the material for paper-making was imported. This was the case even with paper made from rags, for which in the late 1840s British sources supplied less than one-quarter of the requirements, most of the imports coming from Italy and Germany (*National Cyclopaedia*, 1850.) In the 1860s, esparto grass, mainly from Almeria and Murcia in south-eastern Spain or Oran in North Africa, was first imported on a considerable scale for paper manufacture. In three or four years to 1866 the trade grew to a total of 160,000 tons, and by the late 1880s imports had reached a quarter of a million tons annually. It came as cargo in vessels which had previously returned in ballast after carrying coal and machinery to Spanish mining districts. Initially the esparto was not baled and so had to come in bulk to the few British unloading points which also took Spanish minerals that provided the necessary ballast for such a light cargo. The Tyne was the leading destination. Any esparto grass not used there was distributed by inland transport. Later, when baling equipment was installed at the main Mediterranean ports of export for esparto, the points of importation into the United Kingdom became more widespread. (*J.S.S.* 29, 1866; *C.T.J.*, 17 Aug. 1889, p.97.) Subsequently, wood pulp, bleached not by alkali but by sulphuric acid, became important for poorer quality papers. By 1889–90 there were plants in Hull and Barrow using this process and, on the north-east coast, works at West Hartlepool and Marsden between South Shields and Sunderland. From 1891 to 1905 imports of wood pulp into the United Kingdom increased from 157,000 tons to 576,000 tons. (*C.T.J.*, 21 Dec. 1889, p.387, 4 Jan. 1890, p.2, 5 Apr. 1890, p.226, and Macrosty, 1907, p.191.) Recovery of caustic soda at the paper-mills, and the establishment by some of them of their own caustic soda plants lessened the importance of this market for the alkali works at the turn of the century.

As with soap and glass, Merseyside and Tyneside were early leaders in the paper trade. In the course of the nineteenth century growth focused increasingly on the former, to the serious detriment of Tyneside. In 1835 the Newcastle district made most of the second-class paper in the United Kingdom, ranking above Manchester and South Lancashire. There were several large paper-mills in Newcastle. (Fullarton, 1842.) In first-class paper Tyneside was second only to the Manchester district, though its output was only a little more than one-third as great. (Leifchild, 1843, p.254.)

The final, major, and most rapidly growing outlet for alkali was the textile industry; estimates only are available for details of the tonnages used but they do show a much more rapid increase in consumption than in glass or soap. Clearly the North-East and even Scottish centres of alkali manufacture were much less favoured in relation to this trade than those of the Mersey basin (Table 3).

Table 3. *Estimated alkali requirements of leading British industries, 1814–34, in terms of 100% sodium carbonate* (thousand tons)

	Cotton	*Hard soap*	*Glass*
1814	0.72	3.86	4.34
1820	2.20	4.40	5.50
1825	2.92	5.50	5.44
1830	3.90	6.29	5.26
1834	4.46	7.73	4.81

Source: Barker, Dickinson, & Hardie, 1972, p.144.

Overseas trade

Britain dominated world production and trade in alkali in the mid-nineteenth century. Exports to western and central Europe and to the Russian Empire fell away after the 1870s as, with tariff protection and new processes of manufacture, these countries built up their own industry. Shipment to the United States began about 1840 and eventually became far and away the major export market of the United Kingdom industry. The United States never had a securely established Leblanc alkali industry, but after the mid-1880s it began to build up a Solvay alkali capacity. In the 1890s more effective tariff protection, and in the mid-nineties electrolytic manufacture for caustic soda and bleaching powder, brought rapid erosion of the position of the British makers in the United States markets. After this, even though Britain still dominated in the Empire and in neutral markets in the developing world, decline was inevitable. The maximum annual alkali export of 347,000 tons was reached in 1883; by 1900 shipments were already down to 183,000 tons.

CHAPTER 2

An Eotechnic Industry

Natural soda

The earliest and for centuries the only significant sources of soda were highly saline lake deposits, notably the natron and trona lakes of Egypt. During the nineteenth century similar highly alkaline desert lakes were discovered in northern Mexico and in the western United States. Mono Lake, on the Great Basin edge of California, was enthusiastically described as 'more detergent than soapsuds' (Knight, 1873, p.96). At the end of the century much bigger deposits were found in East Africa. The drawback with all these sources was their inaccessibility. Time and again difficulties of transportation frustrated superficially attractive proposals to work them in competition with costly, but large-scale and better located manufacturing processes for alkali.

In the middle years of the century, prominent men of Sonora State, Mexico, tried to work a large soda deposit west of El Altar in the north-west. Transport problems proved too great, and the project collapsed when the vessel carrying the first shipment sank (*C.T.J.*, 17 Apr. 1897, p.266). Attempts to open United States deposits were of no real significance. The modern history of the Egyptian deposits is still more conclusive. Although distances were much shorter, here too transport proved the key problem. By the early 1880s the traveller on the Cairo to Alexandria railway could see at Kafr Dawar station large heaps of natron containing of the order of 60 per cent soda which had been carried in by camel to the extent of 500 tons a year. For a time the Egyptian government continued a generally unprofitable production, but at the end of the 1890s a private company, the Egyptian Salt and Soda Company, was formed to work the natron. It built a modern factory, planned a railway link, and set itself the apparently modest target of competing with British alkali in the Alexandrian market. (*J.S.C.I.*, 28 Feb. 1883, p.116, 30 Sept. 1898, p.889, and *C.T.J.*, 11 Nov. 1899, pp.313–14.) But the works was 35 miles out in the desert, 85 miles from Alexandria, and the only fuel (needed to calcine the natron) was rushes, whose equivalent coal value was 12*s*. a ton of soda ash. An outside expert put costs of production at £2. 10*s*. 0*d*. a ton even with good plant. To this had to be added another 10*s*. a ton for packing, £1. 0*s*. 0*d*. for carriage to Alexandria, and a royalty to the Egyptian government of £1. 0*s*. 0*d*.—in short, delivered costs of £5. 0*s*. 0*d*. a ton. At that time European soda ash c.i.f. Alexandria was £4. 5*s*. 0*d*. (*C.T.J.*, 11 Nov. 1899, p.319.)

The deposit of natural soda at Lake Magadi in the rift valley in southern Kenya was first vaguely known in the early 1870s. In 1902 an East African

Syndicate expedition examined the possibility of working it. There was dispute as to whether it should be calcined at Mombasa or on the Thames. In 1911 the Magadi Soda Company was formed, and trading began in 1915 against the opposition of established soda ash manufacturers. Wartime conditions interrupted trade, and the company was liquidated in 1923. In 1924, in order to control rather than continue to compete with Magadi Soda in the increasingly important Far Eastern markets, Brunner Mond absorbed the company. Magadi remains a small producer. (Hill, 1964, *passim*; *Standard Bank Review*, Nov. 1966, pp.2 and 3; and I.C.I. Records, London, Box 12.1.7A, and *Record* 1926, p.21.)

Vegetation-based alkali production

As Western Europe became increasingly industrialized in the course of the eighteenth century, a widening range of sources of supply of both potash and soda salts had to be exploited. Lees (the waste liquids from hard soap manufacture) and the sands discarded by glassworks yielded alkali, but this could not satisfy expanding needs. British consuming trades were prominent in extending the field of procurement so that from 'forests in Eastern Europe and in the British colonies, manufacturers in Leith, Glasgow, Newcastle and Liverpool imported considerable quantities of wood ash (potash or pearl ash [an impure pottasium carbonate]). From the Mediterranean came the ash of Salsola Soda (barilla) and from the shores of Scotland, Ireland and Norway seaweed ash (kelp).' (Clow, 1947, pp.297–316.) Two reports of 1765 illustrate the growing demand and the widening supply patterns:

> A quantity of pearl ashes, the manufacture of Minorca, have lately been imported, and is esteemed little inferior to those brought from Trieste in Italy.
>
> The inhabitants of New England, have at last luckily hit on the process of making potashes, and of such a strong alkali, as must soon outvie the Hungary sort for making glass and bleaching; and the great advantages derived from this manufacture, not only in point of price, but in the clearing of their woods, and the manuring of their land, has made the land holders, one and all, to set about making them. (*Annual Register*, 1765, pp.115 and 138.)

A decade later the disturbances of the American Revolution interfered with this trade. One result was to encourage search for a practical process for alkali manufacture from a mineral base. Another was the extension of trade in barilla and in kelp. In each area of alkali consumption, the supply in part determined which source of alkali should be used, but the choice was also affected by the district's particular trades. For instance, in the 1830s and even into the 1840s, barilla, kelp, and wood ashes were all used for inferior kinds of glass—kelp commonly for common green glass, but purified pearl ash for crown and fine flint grades. (*Penny Cyclopaedia*, 1835; *National Cyclopaedia*, 1849.)

Barilla

While the ashes of land plants mostly yield potash salts, those of marine plants provide a larger quantity of sodium salts. Barilla was the commercial name for the impure sodium carbonate produced from marine plants. It was derived from coastal grasses and was imported from the Mediterranean, notably from Alicante but also from Malaga and Carthagena, from Teneriffe and the neighbourhood of Narbonne and Aigues-Mortes in the South of France. (Richardson & Watts, 1863, pp.193–5.) Commonly the plants were gathered in September and dried in the sun. In October they were burned in a hemispherical kiln dug in the earth. The soda contained by the plant became fused into masses of as much as one and a half tons in each kiln. This barilla often contained no more than 16–24 per cent of sodium carbonate, though sometimes much more. When broken into fragments it was ready for shipment. The best barilla came from Alicante, which in good times shipped as much as 90,000 cwt a year, and itself became an important centre of the Spanish soap trade.

At the end of the Napoleonic Wars one London soap-maker, T. and B. Hawes, employed about 1,000 tons of barilla and 500 tons of kelp annually. (Salt Duties 1818, pp.362 and 496.) In the twenties import duties on barilla were gradually reduced, from 11*s*. 4*d*. per cwt until 1822, to 8*s*. 6*d*., and then in 1831 to 2*s*. per cwt. The trade reached new high levels, the average importation between 1829 and 1834 being 252,000 cwt (Table 4). Only a few years later imports were shrinking, and by 1843 were less than 52,000 cwt. By then the annual British consumption of sodium carbonate was seven times as great as the record level of barilla importation. The mineral-based alkali industry had by this time become predominant. Before that, however, another, indigenous, vegetation-based source of supply had been of great importance. (*National Cyclopaedia*, 1849; Knight, 1851, p.309.)

Table 4. *Barilla imports, 1833* (cwt)

Canary Islands	74,537
Spain and the Balearic Islands	111,748
Italy and Italian islands	24,783
Other	3,422
Total	214,490

Source: *Penny Cyclopaedia* 3, 1835.

Kelp and the problems of raw material export economies

As early as 1800 a duty of £44,000 was paid on imports of pearl ash and potash and of barilla. Kelp was a major domestic source of alkali, obtainable without duty, though inferior to barilla in quality. Kelp is the ash left from the combustion of seaweeds (wracks or tangles of the species *Fucus* or *Laminaria*). In the British Isles they are found mainly on the western and north-western coasts. The industry seems to have begun in Ireland, but soon became important in Scotland. As early as 1662 kelp was being used in glass-making at Prestonpans; some twenty years later it was made at Anstruther across the Firth of Forth. In the Orkneys kelping began in 1722, and within a few years

Orkney kelp was being sold to Tyneside bottle-makers. The Hebrides made kelp at least from 1730. With the increasing demand for alkali, and shortage of wood ash, output grew and prices moved upwards in the 1790s. By this time the western Highlands were large suppliers to Liverpool soap- and glass-works and Glasgow merchants travelled to Mull to buy kelp from the few wealthy proprietors who monopolized the trade. (Clow, 1947; Saint Fond, 1907, p.79.)

Kelping provided a good deal of seasonal employment on the Scottish islands and for crofters on the mainland, and much income for the lairds, but its long-term significance to the Highland economy has been much disputed. When near its peak at the end of the Napoleonic Wars, the trade was described to a Select Committee by Stevenson of Edinburgh (Salt Duties 1818, pp.519f.): annual kelp production of 17,000–25,000 tons provided seasonal employment for anything from 60 to 100,000 people. Generally it provided half the income of Orkney coastal estates; in exceptional cases, land rates of only £300 a year were matched by an income from kelp of over £1,500. Kelp kilns had by this time been built in the Western Isles—the Highland Society had offered big premiums for the design of kilns—and in various ports considerable sums were invested in kelp storehouses. There were wider effects too. Some two hundred vessels of 60–100 tons burthen were seasonally employed: delivering kelp to the Tyne or Wear, they carried coal back to the Highlands (Table 5).

Table 5 *Kelp imports by leading ports, 1818, 1825* (tons)

	1818	*1825*
Liverpool	4,527	4,163
Newcastle	1,852	3,668
Leith	1,997	1,289
Greenock	1,280	621
Glasgow	1,223	112
Hull	488	295

Source: *B.P.P.*, 1829, XVII, p.381.

But the work was hard, and the returns to the worker small. By 1814, for three months' labour a man averaged only one and a half tons of kelp worth about £15. (Clow, 1947, pp.306–7.) Thirty years later, North Uist's annual output of 900 tons required labour from 400 persons in June, July, and August, and the average earnings for each family for this season were as little as £4. (Lewis, 1847.) By the 1860s, when small quantities of a number of special salts—chiefly iodine and bromine—as well as chloride of potassium (5–6 cwt) and sulphate of potassium (3 cwt) were obtained from each ton of kelp, it was reckoned that this required the collection of up to 22 tons of wet seaweed. (Levi, 1865.) The prices for kelp and the return for the hard labour which was involved (sometimes fifteen hours a day) seesawed after that—they seem to have risen again in the late 1880s. (*C.T.J.*, 14 June 1888.) But the return was generally low and uncertain. Edward C. C. Stanford, who took great interest in the chemistry of seaweed and particularly the extraction of iodine, summed up the situation: 'The kelp burner in the good old times only received £2 per ton for his labours; this is 2/- per ton for every ton of seaweed, which has to be cut, brought ashore, dried and burned, and even then all the labour may be lost by wet weather.' (Stanford, 1884, p.298.)

Kelping also diverted labour and attention from other activities. True, at the end of the kelping season the workers usually went on to fishing, but meanwhile they had neglected their farms or crofts. In the *New Statistical Account*, Cameron (1845) described how, in Ross and Cromarty, economy and society were wholly bound up with the seasonal trade:

> Kelp is the principal manufacture in the parish. It is almost impossible to tell how many are employed in the work, for young and old, male and female, all who are able to carry a creel of ware, or help to fill it, are engaged in manufacturing it at different times, for three months, from the time of cutting the ware till the solid mass is weighed on board the vessel. The price of kelp varied from £1. 10*s*. to £3. 3*s*. per ton; at £3. 3*s*. per ton the sum may be as much as the man and wife in one family could earn in three months by any other work; but the toil in cutting, drying, burning the ware and watching the pot day and night till the ware is converted to boiling lava is terrible and would require extraordinary wages. This process if not injurious to health is ruinous to the eyes.

Opportunity costs were high:

> But when thousands are engaged, all the summer season, making kelp their crofts and lots are neglected, potato fields are overrun with weeds, consequently the return is small, and part of the gain in kelping is lost in their potato crop; their cattle are much neglected; cornfields are destroyed; and the tenants are distressed for their rents. Many of the herd boys that should attend the cattle during the summer heats are kelping; many beasts are lost in mossy veins, and fall from rocks when they run wild during an excessively hot day, —so that in this way, the gain by kelp becomes a loss.

Others (notably McCulloch (1841)) confirmed Cameron's impressions.

Kelp prices had generally moved upwards in the second half of the eighteenth century, and at the beginning of the nineteenth the usual price was about £22 a ton. In the twenty-three years to 1822 the average was £10. 9*s*. 7*d*. a ton. From 1822 to 1832 prices fell from £10. 10*s*. to £2; ten years later the average was £3 to £4, but it had been as low as £1. 10*s*. (Clow, 1947; McCulloch, 1841; Goodrich-Freer, 1902.) The Orkneys, the largest producer at the beginning of the century, exported only 500 tons in 1833, and by then kelp was one-thirtieth of the value of all its exports. Demand for Argyllshire kelp was reported as 'much diminished' by 1834. By the early 1840s production at Orphir on Mainland had fallen from its old level of 70 tons to 20 tons, and on Westray kelp was scarcely worth making though it had formerly been 'the principal manufacture and grand staple'. (*Penny Cyclopaedia*, 1834, IV, and *New Statistical Account*, 1845.)

Later demand for iodine was insufficient for a sustained revival of the trade. With the decline of the kelp business went a further shrinkage of the Highland fringe and of island economy and society. On Bara the factory in which the MacNeills had tried to make soap—and, according to some, even glass—was abandoned; only two tall octagonal chimneys remained by the early 1860s. (Stanford, 1884, p.298; Goodrich-Freer, 1902, p.37.)

The cause of this collapse was in part the liberalizing of the trade in barilla, but much more decisive was the repeal of the Salt Duties which permitted the growth of a mineral-based soda industry. The free trader McCulloch approved of the 'more liberal line of commercial policy' which by the late 1830s had in some cases already led to seawrack being collected only to manure the land or to supply winter cattle feed in hard seasons (McCulloch, 1841; *Penny Cyclopaedia*, XIII, 1839, article 'Kelp'), but others inveighed against it. In Shetland, even though kelp had never been as important as in the Orkneys or Hebrides, the want of income from it was seriously felt by the early 1840s.

> It is difficult to conjecture what other motive, than a reckless adherence to the speculative dogma of Free Trade could have induced the British Legislature to inflict so severe a wound on the industry and resources of the Scottish islands and which neither equity nor policy can honestly defend; much more oppressive and pernicious monopolies may take shelter under the mask of commercial freedom, than under the protective mantle of prudent and patriotic government. Free Trade is, after all, only relative; it has been imagined and described; but when the globe shall become one vast Utopia, then only can we expect that its *beau ideal* will be realized. (Edmondston, 1845.)

These bold and prophetic words did not attract much attention in the high-tide phase of free trade capitalism. Others noted the more direct relationship with the Leblanc process: 'Soda ash, highland crofters, Irish peasants—they are all of a piece like the moons and tides. Leblanc's invention has had not a little to do with the Irish Land Question and the Crofter Difficulty.' (*C.T.J.* 7 Dec. 1889, Trade Notes.) Leblanc's existence 'has been as great a misfortune to the Outer Hebrides as if he had been a modern landlord' was Goodrich-Freer's (1902) even more trenchant summing-up. Strangely enough, eighty years earlier, the Salt Duties Committee had recognized the regional impact of Salt Duty repeal and had faced in some measure the balancing of localized gain against loss in other districts. They observed that a practicable process for making alkali from salt might not only end the need to import barilla but, by displacing kelp, 'compel a large proportion of the population of the North West of Scotland to the necessity of emigration'. This was, they stressed, still conjectural for 'no process has been specified to Your Committee by which alkali could be profitably produced by the decomposition of salt; yet as such a process may exist or be hereafter discovered, they have thought it their duty to enquire how far in its consequences it might be one of public benefit, particularly as they have reason to believe considerable alarm to have been entertained in Scotland in reference to its eventual effects on the kelp trade'. (Salt Duties 1818, pp.343 and 344.) In fact, twenty-seven years before, in September 1791, a patent had been granted for a practicable process of salt-based alkali manufacture. In 1823 the Salt Duty was sharply reduced and in 1825 repealed. Salt for chemical purposes cost £19 a ton in Glasgow in 1809, but by 1824 was already as little as £2. 6*s*. (Mactear, 1877, p.5.) The way was now clear for the growth of an industry which would sweep away employment for thousands of crofters, Mediterranean barilla-makers, and the casual potash sales of woodsmen and temperate-zone farmers.

CHAPTER 3

Processes in Mineral-based Alkali Manufacture

Mineral processes of alkali manufacture depend upon common salt either dry or as brine. From it are derived the soda ash, soda crystals, bicarbonate of soda, caustic soda, bleaching powder, and the multitude of minor products which made up the classical alkali industry. Only three manufacturing processes have been of practical significance—the Leblanc, the ammonia soda, and the electrolytic. Leblanc's process was the most roundabout, the most wasteful and laborious but capable of the widest range of products. It dominated world production of alkali in the two middle quarters of the nineteenth century. The ammonia soda process was not a practical success until 1862 and by the late 1880s had gained a pre-eminence for soda ash which it has never lost. Electrolytic processes were limited to chlorine and caustic soda and were not practicable until 1890s (Vincent, *c.* 1877, pp.787–847; Lunge, 1910).

Leblanc's process

Leblanc's process remained virtually unchanged throughout its eighty years of dominance in the alkali trade. Initially it was very wasteful and its industrial history was marked by a number of innovations which greatly reduced costs of production. Henry Louis Duhamel seems to have been the first to recognize the distinction between soda and potash salts in 1736. Increasing industrial demand and the loss of some sources of natural ash supply led to shortages and rising prices so that strenuous attempts were made to produce sodium and potassium carbonate from common salt. In 1781 Keir of Tipton went so far as to produce enough soda from sea-salt to supply the needs of his soap-works. (Hardie & Pratt, 1966, p.19.) Later the Earl of Dundonald in the North of England and the Swedish chemist Karl Scheele made some important experiments, but progress was concentrated primarily in France. In 1775 the French Academy had offered a prize for a salt-based alkali process. Malherbe and De la Métherie made significant discoveries, but laboratory achievement was not followed by successful manufacture and none of these pioneer processes proved able to compete with soda derived from seaplants. (Gillespie, 1957, p.158.) In 1787 De la Métherie's record of his experiments was read by Nicholas Leblanc, surgeon to the Duke of Orleans, who added chalk to his predecessor's mixture of sulphate and charcoal, and from this succeeded in producing sodium carbonate. Leblanc's process was essentially the Black Ash process. He worked on the production of sodium sulphate (saltcake), but his methods in that respect were clumsy. (Gillespie, 1957, p.155; Reilly, 1951, pp.288–90.) In 1791, with financial assistance and the protection of a patent for fifteen years, Leblanc built works for his process at St. Denis. But revolutionary disturbance was followed by state control of the works, and even after their eventual restoration to him in 1800 they never operated with success. Leblanc suffered a breakdown and in 1806 committed suicide.

The Leblanc process has two distinct stages. The first is the manufacture of saltcake or sodium sulphate. Saltcake is then fluxed with calcium carbonate and carbon (charcoal in Leblanc's original scheme, but coal in normal Leblanc practice) to produce black ash, a mixture of calcium sulphate and sodium carbonate but with a number of other impurities. Sodium carbonate is produced from the black ash by 'lixiviation', that is it is dissolved by water and then evaporated to produce the solid soda ash. The two essential chemical processes are:

(1) $2NaCl + H_2SO_4 = Na_2SO_4 + 2HCl$;

(2) $Na_2SO_4 + CaCO_3 + 2C = 2CO_2 + Na_2CO_3 + CaS$

The manufacture of sulphuric acid and saltcake

As sulphuric acid was required in considerable tonnages and was difficult to transport, most alkali-makers produced their own. The acid was originally prepared by heating sulphates—alum, copperas (iron sulphate often obtained from the coal measures)—and condensing the products of distillation. It was a notable improvement when in the first half of the eighteenth century Ward of Twickenham introduced the Bell Chamber method. In this, sulphur and nitre were burned, the vapour was absorbed in water, and the solution was concentrated. About 1746, John Roebuck of Birmingham, while using essentially the same process, moved on to operations on an even larger scale by replacing the glass vessels with lead boxes (lead chambers, as they became commonly called). The base of each chamber was covered with a few inches of water, the sulphur and nitre being carried on a tray.

During the nineteenth century Roebuck's lead-chamber process was conducted on a progressively bigger scale and by the early twentieth century chambers were commonly 100–200 ft long, 20–30 ft wide, and 15–25 ft high. Acid manufacture was also made into a continuous rather than an intermittent process. Commercially successful production required both that the nitrous fumes originally carried away into the atmosphere should be recovered and that the nitre and sulphuric acid from this first recovery process should be fully re-employed in the lead chambers.

Measures to avoid wastage involved the addition of the tall Gay Lussac tower (from 1835) and the shorter Glover tower (introduced in 1859) to the sulphuric acid plant. Works varied greatly in size—in the seventies a small one might make 150–200 tons of acid a year and a big one 4,000–5,000 tons. (Kingzett, 1877, p.49.) An alternative process, the 'contact' method of sulphuric acid manufacture, though experimented with seventy years earlier, did not become practicable until the very end of the nineteenth century. Until World War I almost all British sulphuric acid capacity used the lead-chamber method. (Sulphuric Acid, 1918, p.3.)

Stage one of the Leblanc alkali process proper involved the 'decomposition' of common salt by sulphuric acid in cast-iron pans or pots known as saltcake or decomposing furnaces. In the late nineteenth century commonly one half a ton of salt and a half ton of sulphuric acid were worked together in a large hemispherical cast-iron pan. (Partington, 1919, p.63.) The product of the saltcake furnaces was sulphate of soda, used directly by glassworks or, in its purified form, sold as the purgative Glauber's salts. The vigorous reaction involved also produced large quantities of hydrochloric acid gas (muriatic gas). This made work in the saltcake process highly disagreeable, and eventually mechanical furnaces were devised in which the charge was stirred by machinery. The first mechanical furnace, made by Jones & Walsh, was patented in 1875; but the most successful was introduced by James Mactear at the St. Rollox works, Glasgow. Apart from its high capital cost, one of the main disadvantages of Mactear's furnace was that the hydrochloric acid gas became mixed with fire gases and was therefore available in a less pure form than that from manually operated saltcake furnaces. Recovery of hydrochloric acid gas had been in itself one of the impressive material advances in the Leblanc process.

In the early decades of the process, hydrochloric acid gas was simply emitted into the atmosphere. The result was disastrous for metalware, building materials, natural vegetation, and crops, and led to extensive litigation which was not only a significant cost factor but also affected the location of the industry. Continental works eventually solved the problem by passing the gas from the saltcake furnaces through a long series of Woulfe's bottles in the form of stoneware receivers, in which most of the gas was washed out; the final stage was achieved by means of a small a tower packed with coke in which the remaining gas was dissolved in a downward trickle of water. In Britain, where there was less concern to recover acid, the small tower was extended into a tall tower, essentially Gay Lussac's coke tower but adapted by William Gossage for hydrochloric acid gas recovery. The Alkali Act of 1863 required that not less than 95 per cent of the hydrochloric acid gas involved should be absorbed. In fact, by means of the Gossage tower, 99 per cent recovery rates were common by the 1870s, and by 1887, partly owing to the opportunities for economy but largely owing to the careful supervision of the alkali inspectors, hydrochloric acid gas liberated into the atmosphere amounted to no more than 0.1 per cent of the total gas produced (Kingzett, 1877, p.86; Fletcher, 1887, p.326). Looking back, one of the industry's technical experts observed, 'There is hardly another example in the annals of legislative efforts equal to this, in respect of the real benefit conferred by it both on the general public and on the manufacturers themselves.' (Lunge, 1910.) Their by-product output of hydrochloric acid led more alkali-makers into large-scale production of chlorine and of bleaching powder, already important factors in the growth of the trade.

Chlorine and bleaching powder

Chlorine production had involved a mixture of common salt, sulphuric acid, and manganese peroxide in a spherical leaden vessel. With the availability of recovered hydrochloric acid in the Leblanc process a new method was introduced. Hydrochloric acid was oxidized by the use of manganese dioxide in 'chlorine stills' made either of stoneware or of acid-proof rock. The process was wasteful, and therefore costly, as much of the acid and all the manganese was lost. It was also most disagreeable as the 'still liquor' that had to be disposed of contained both free hydrochloric acid and free chlorine which caused great nuisance in watercourses and drains. Moreover, as the price of bleaching powder fell, such waste could not be tolerated. The Weldon process, first applied in 1866, involved the treating of this obnoxious still liquor with air, chalk and limestone. The regeneration of the manganese dioxide was thus achieved with only a very small loss of manganese. Unfortunately much of the hydrochloric acid was still lost, though now as the solid waste calcium chloride. The Lancashire alkali-maker Henry Deacon, working with Ferdinand Hurter, introduced a cheaper and more direct process in 1868, producing chlorine gas and water directly by passing hydrochloric acid gas and air over hot brickwork impregnated with a catalyst, copper chloride. This process had the disadvantage when compared with Weldon's of requiring much more costly and complicated equipment which needed a great deal of attention. As a result, in spite of its deficiencies, right up to the early years of the twentieth century most of the world's chlorine was made by Weldon's process. Like Glover's denitrating tower, but on a far greater scale, Weldon's process cut the over-all cost of the Leblanc trade (Table 6).

Table 6. *Comparative costs of chlorine manufacture by the traditional and Weldon processes*

	£	s.	d.
Native manganese process			
16 cwt of manganese at 145*s*. a ton	5	16	0
14 cwt of lime at 16*s*.		11	2
Steam		5	0
Labour		12	6
	7	4	8
Weldon's process			
25 cwt of artificial manganite at 12*s*.		15	0
27 cwt of lime at 16*s*. a ton	1	1	7
4 cwt of limestone at 12*s*.		2	5
Labour		10	0
	2	9	0

Source: *M.J.*, 13 Feb. 1875, p.162.

Though it had a wide variety of uses, most chlorine was used in bleaching. In 1775 Berthollet had first used chlorine as a bleaching agent. In the late 1790s it was superseded by the more conveniently handled and cheaper bleaching powder introduced by Charles Tennant. Bleaching powder was produced by the action of chlorine on hydrated lime. With hydrochloric acid gas recovery, chlorine manufacture, and the production of bleaching powder, what had formerly been a branch rapidly became the main stem of the Leblanc industry, and indeed in the last decades of the nineteenth century, the process of manufacture tended increasingly to stop at this point, the associated sulphate of soda, now virtually a by-product, being sold direct to the glass-makers who used it in place of soda ash. In the twentieth century, development of chlorine sales in cylinders or tank cars gradually helped it to supersede bleaching powder as a bleaching agent, essentially a reversion to the eighteenth-century processes though with different techniques of transport.

Soda ash manufacture from saltcake

The production of carbonate of soda, soda ash, the process in which Leblanc's innovations were decisive, involved the further processing of saltcake in a reverberatory furnace known as a 'balling furnace'. Sodium sulphate was mixed with lime and carbon and heated. The product, black ash, was an impure carbonate mixed with sodium sulphide and unburnt coal, with a residue composed largely of calcium sulphide. The proportions used in this manufacturing process varied widely—in the eighties for each 100 parts of saltcake there were from 100 to 140 parts of chalk or limestone and 40–70 parts of coal. Later the proportions seem to have been 100 saltcake, 100 limestone, and as little as 35.5 coal. (Note Partington, 1919, pp.63–71.) While exposed to the flame for charge times of from 40 to 50 minutes the mixture had to be constantly turned over by a 'paddle' or a rake by the 'ball furnace man', a job involving not only considerable judgement and experience but extremely hard physical labour. Before the middle of the century attempts were made to introduce mechanical balling. In 1848 W. W. Pattinson produced a rotating ball furnace, but owing to severe wear and tear this was not successful. Eventually success was attained in another revolving furnace commonly known as 'the revolver'. The sodium carbonate was broken up and lixiviated from the black ash with warm water and then concentrated by evaporation to produce commercially pure soda ash. The insoluble residue, variously known as 'alkali waste', 'tank waste', 'vat waste', or in Staffordshire 'bank billy', had to be dumped around the works, where it caused a great loss of space and poisoned air and water. The loss of sulphur was on a very large scale (from 20 to 25 per cent of the waste being sulphur) and sometimes amounted to over half the total cost of the soda ash. Nationally the annual sulphur loss was as much as 150,000 tons, worth £0.4 million.

On grounds both of amenity and of works economy there was a great incentive to recover sulphur from solid waste. This was attempted from 1837 onwards, and in 1861 Schaffner, and a year later Mond, made more successful innovations in this direction. The decisive breakthrough, however, came only in the eighties as the result of work by Claus (1882) and Chance (1888). Chance used flue gas which was passed through alkali waste suspended in water; the sulphur being recovered at the next stage in a kiln operation. His experiments in the early eighties were embarrassed by a fall in the price of pyrites which reduced the price of sulphur by 50 per cent, but he persevered and carried the matter to success a few years later. (Fletcher, 1887.) Eighty-five per cent of the sulphur could be recovered by this process and the sulphur produced was of a high degree of purity. The cost competitiveness of the Leblanc process was consequently extended. However, even though the cramped space of British Leblanc alkali works encouraged them to work up their waste, trade uncertainties slowed the advance of sulphur recovery. (*J.S.C.I.*, 28 Feb. 1890, p.182; *C.T.J.*, 4 Nov. 1893, p.293.)

Soda ash was converted into washing soda for domestic use by dissolving the ash in hot water and then leaving it to cool and crystallize, a process which took about a week in winter and as much as twice as long in summer. Bicarbonate of soda was derived from soda crystals by the action of carbonic acid. By the end of the century bicarbonate, partly for medicine but mostly as baking powder, was commonly made by the rival ammonia soda process, though for a time it retained an ammoniacal smell which lessened its attractiveness. As the ammonia soda process came to dominate also the production of soda ash, eventually most of the Leblanc liquor from the black ash process was converted into caustic soda, involving the action of slaked lime on a diluted solution of soda ash. Many other chemicals were produced by the Leblanc works in smaller tonnages, for instance hyposulphite of soda was made from sodium carbonate and sulphur for the removal of the last traces of chlorine from wood pulp and for photographic purposes.

The Leblanc process permitted the production of a great number of joint and by-products but it was complicated and slow. Yet its complications provided many opportunities for economy and for new finished products. Gay Lussac's 1844 method made it possible to save over 50 per cent of the nitre used, but in the early 1860s, with nitrate of soda at £12 a ton, it did 'not pay to collect and absorb the waste oxides of nitrogen' (Schunk *et al.*, 1861, p.156). Glover's denitrating tower cut nitre consumption from 90 lb to 15 lb a ton, and Weldon's manganese process reduced the use of manganese for every ton of bleaching powder from 16 cwt to about 30 lb by the late 1880s. (Stuart, 1889, pp.145–6.) This adaptability exemplifies both the weaknesses of the Leblanc process and its ability to struggle on for so long in competition with the far simpler, chemically more elegant, and commercially more efficient ammonia soda process (Fig.1).

The ammonia soda process

The ammonia soda process involves the mutual decomposition of ammonium

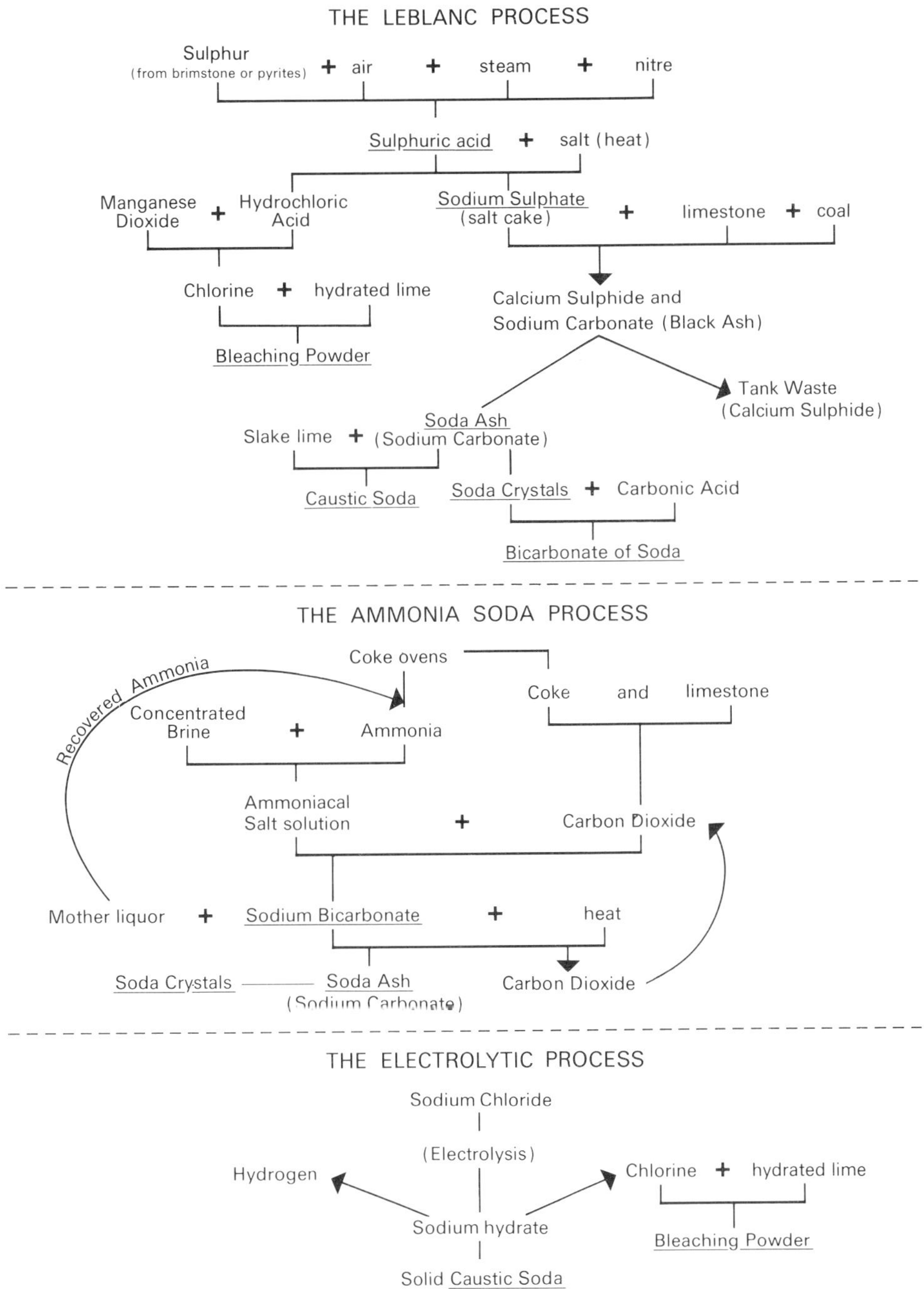

Major saleable items are underlined in each process.

Fig. 1. Flow diagram of Leblanc, ammonia soda, and electrolytic processes

bicarbonate and sodium chloride to produce sodium bicarbonate which is then converted into soda ash by the application of heat. The two essential reactions are

$$NaCl + NH_4HCO_3 = NaHCO_3 + NH_4Cl;$$

$$2Na\,HCO_3 = Na_2CO_3 + CO_2 + H_2O.$$

The ammonia soda process was known theoretically as early as 1811, if not, hazily at least, as early as Leblanc's own innovation. (Reilly, 1951, p.291.) It was patented in England by Dyar and Hemming in 1838 and carried out by them in that year on an experimental scale in Whitechapel. There were at least eight ammonia soda patents between 1838 and the early 1860s. (Richardson & Watts, 1863, pp.216, 217, and 227–31.) All proved commercial failures. The main reasons were imperfect conversion of the salt and, more important, the loss of ammonia, a problem of which Dyar and Hemming were well aware and for which they tried to make provision. A commercially successful ammonia recovery apparatus was at last introduced by Ernest Solvay in 1861. It was in its Solvay variant that the ammonia soda process came to dominate the world's alkali trade in the course of the next thirty years. Perhaps as critical to its success as Solvay's own innovation were those introduced by Ludwig Mond in struggling to make the process commercially successful in Cheshire.

Solvay's process begins with brine. Although the brine has to be concentrated and purified, in this form the sodium chloride input was cheap in comparison with the dry salt used in Leblanc works. The dense brine solution is saturated with ammonia, the resulting ammoniacal salt solution being further saturated with carbon dioxide in the form of gases made in kilns charged with limestone and with coke. The carbonizing of the ammoniacal solution takes place in the so-called Solvay tower in which at numerous levels the gas stream is divided and brought into intimate contact with the solution. Great care is necessary to maintain a constant temperature, for if too cool the resulting sodium bicarbonate has so fine a crystalline structure that it is difficult to filter, whereas if too warm the reaction is reversed and sodium bicarbonate already formed breaks down into sodium chloride and ammonium bicarbonate again. By periodic withdrawal of part of the solution from the Solvay tower and its replacement by a fresh solution of ammoniacal salt the process is made continuous. The bicarbonate is then heated to produce soda ash and carbon dioxide, the latter of which is recycled.

The ammonia soda process required more capital outlay than the Leblanc process, but it involved less fuel, no sulphur, and used inexpensive brine rather than salt. By 1884 it was estimated that Leblanc soda ash cost 70 per cent more than Solvay ash. (*M.J.*, 19 July 1884, p.852.) But, as had been suggested by earlier experiments, critical to the commercial success of the new process was the ability to recycle ammonia. (Kingzett, 1877, pp.164, 165.) Ammonia was at first provided from gasworks liquors but the supply was later increased by installation of by-product coke ovens, Solvay's work on this involving the scrubbing of by-product gases and leading to the development of the Semet–Solvay oven design. Ammonia recovery was achieved also at the raw coal-using blast furnaces in Scotland. The loss of ammonia in the Solvay process was almost halved between 1862 and 1887, mainly as a result of Mond's work. (*C.N.*, 17 Dec. 1907, pp.303, 304.) Later, ammonia was largely recovered from the mother liquors left after the vacuum filtering of the muddy sodium bicarbonate from the Solvay tower. However, weaknesses of the Solvay process remained the loss of salt and of all the chlorine contained in the salt as either sodium chloride or calcium chloride, the latter being dumped as solid waste, though on a very much smaller scale than that of the tank waste of the Leblanc process. In 1896 Mond claimed that he had succeeded in recovering chlorine from the waste and that now therefore the ammonia soda process could compete with the Leblanc process in the production of bleaching powder, using lime from the kilns whose prime purpose had been to produce carbonic acid gas for the manufacture of the powder. Commercially, however, this development was frustrated by the high costs of the processes which were tried, and, after the early 1890s, by the success of electrolytic processes in the cheap production of chlorine. (Mond, 1896, p.591.)

Electrolytic alkali

Electrolysis as a means of production of alkaline chlorides was patented as early as 1851. However, as with ammonia soda, laboratory success was slowly converted into commercial viability. Reduction in costs of electricity was the vital factor and it was not until the mid-1890s that electrolytic alkali manufacture became important. Even then its competition was selective, overlapping only slightly with the Solvay process and covering commercially only a part of the vast product range of the Leblanc process. The latter, however, proved a critical part. In the electrolysis of a brine solution, hydrogen is produced at the cathode and chlorine at the anode. By this simple process electrolysis came to dominate the production of chlorine and bleaching powder.

$$NaCl = Na + Cl.$$
$$Na + H_2O = NaOH + H.$$

If the hydrogen escapes and the chlorine is collected separately so that it does not react with the sodium hydrate, the latter can either be sold as such or converted into solid caustic soda, another direct point of competition with the realigned Leblanc industry of the 1890s. Commercially the production level for electrolytic caustic soda was determined by the amount of chlorine which could be sold and therefore largely by the demand for bleaching powder. Electrolysis produced little more than half as much alkali (calculated as soda ash) as bleaching powder and, as, around 1900, world consumption of bleaching powder was at best one-fifth of the tonnage of alkali consumed, not more than one-tenth of all the alkali needed could be provided by electrolysis, though the significance of electrolytic processes in caustic soda were very much greater. The other nine-tenths had to be supplied by other processes. In spite of extended uses for chlorine in the twentieth century, this limitation still applies, though to a smaller extent, so that most of the world's caustic soda is even now made chemically (that is by the action of slaked lime on soda ash) rather than electrolytically. In practice what happened was that the Leblanc process, technically capable of producing all alkali products, was squeezed out on one side by the ammonia soda process, which took over the bulk manufacture of soda ash, soda crystals, and bicarbonate of soda, with some manufacture of caustic soda by the mid-1890s, while on the other hand electrolytic methods came to dominate in bleaching powder and in caustic soda. In spite of its amazing versatility, the elimination of the remnants of the Leblanc process was inevitable. In the final stage the Leblanc firms came to be concerned primarily with sulphur derivatives.

CHAPTER 4

Mineral Raw Materials for the Alkali Industry

Sulphur

The growth of the alkali industry required the assembly and processing of large tonnages of coal, salt, limestone, and sulphur. Sulphuric acid manufacture was an almost universal activity of the alkali firms. The sulphur trade was initially wholly dependent upon imported supplies of brimstone. In the mid-nineteenth century, important sources of sulphur were known at Radoboy in Croatia, at Szwoswice, south of Cracow, at Czarkow, the Isle of Milo, in Murcia, near Apt in Vaucluse, and in the Red Sea, but Sicily was the only source of commercial importance to the early alkali trade. Sicilian sulphur exports to Britain kept pace with the early boom in production of alkali, rising from under 5,000 tons in 1820 to 12,000 in 1830 and almost 45,000 tons in 1838. In 1834/5 Tennant of Glasgow and the Muspratts of Lancashire bought Sicilian mines. In 1838 the king of the Two Sicilies was persuaded to grant a monopoly of the trade to the house of Taix in Marseilles. Prices rose sharply, and in the year to October 1839 only 22,000 tons of sulphur were imported to Britain. By early 1840 the delivered price of sulphur in Britain was £13–£14 a ton as compared with an average of £6. 10*s*.–£7 in 1836 and 1837. British interests in the island and British shipping also suffered. When no redress was forthcoming, the Royal Navy interfered with Neapolitan commerce until in July the monopoly was dissolved.

The Sicilian deposits are embedded in limestone and clay marl and are scattered through the provinces of Caltanicetta, Girgenti, and Catania with two isolated belts at Lercara in Palermo province and Gibellina in Trapani province. Mining and extraction were primitive: into the third quarter of the nineteenth century boys of 8–10 years of age were used to carry the ore to the shafts and up to the surface. Even in the early 1870s, the 'calcaroni', or furnaces, involved 'a simple fusion, where the sulphur itself serves for fuel, and which is carried on in the open air without any regard for waste. Thus the ore which contains fifteen to forty percent of sulphur yields only ten to twenty-five percent.' The high price of fuel—a national problem which, it was suggested at the time, would prevent Italy from ever building up an important chemical industry—frustrated attempts to improve methods of extraction. (*C.N.*, 6 Nov. 1874, p.218; *Encyclopaedia Britannica*, 1887, article 'Sulphur'.) In the mining districts there grew up a large dependent population—some 27,000 were engaged in the operations in the mid-1880s—badly affected in health by sulphur fumes and suffering also from the impact of the fumes on their farming. In the trade crisis of 1839/40 almost the whole population of the sulphur districts was thrown out of employment, 'multitudes of them have become desperate robbers who have rendered the country so insecure that it has been necessary to place it under military law. Persons cannot travel without an escort, and the number of arrests and executions has continued very great. . . .' (Rawson, 1839/40, pp.446–57; Goodwin, 1842.) Transport inadequacies in the interior of Sicily were at last made good by the provision of rail links through the sulphur districts, delivering to four ports which eventually monopolized the shipping—Palermo, Girgenti, Licata and Catania. Extraction methods, however, remained primitive; even in the late eighties miners were said to have to climb 300–500 feet from the bottom of the pit by circuitous steps to bring the ore to the surface on their shoulders. The year 1890 brought business failures, but when prices rose again steam power and mining machinery were introduced to many mines for the first time. (*C.T.J.*, 27 Aug. 1892, p.130.) After the dissolution of the monopoly of 1838–40, Britain resumed importation, but was never wholly dependent upon Sicily again. After about 1851 prices rose yet again and demand increased for use in treating vine disease. By 1871 total Sicilian sulphur exports were 171,000 tons as compared with only 38,000 in 1830, but the tonnage to the United Kingdom was no greater than in the mid-thirties. At this time, while exports to the United Kingdom were falling, the trade to the United States had been increasing (Table 7).

Table 7. *Sicilian sulphur exports, 1871* (thousand tons)

To	
United Kingdom	31.0
France	16.9
United States	41.7
Others	81.5
Total	171.1

Source: *C.N.*, 6 Nov. 1876, p.218.

Late in the nineteenth century, the problems of an antiquated and much divided Sicilian industry, with over-production by expensive mining processes in a setting of generally falling prices, were alleviated for a few years by the formation of an Anglo-Italian sulphur syndicate. Apart from a few thousand tons exported from Japan, Sicily was still effectively the world's only sulphur producer. However, in 1891 Herman Frasch's superheated steam extraction method was devised, and a few years later it was applied to the huge sulphur deposits of Louisiana regarded as unworkable since their discovery more than twenty years earlier. The Frasch process provided increasingly keen competition for Sicily. Long before this, however, the sulphuric acid industry had found new materials.

As soon as the Sicilian monopoly was imposed in 1838, extended attempts were made to use pyrites as an alternative source of sulphur. For the next sixteen years the supply of this increasingly important mineral depended on three home sources: the coal measures, the south-west of England, and pyrites from the Wicklow Mountains. Even when well cleaned before processing, coal measure pyrites always contained some organic matter so that by the late 1860s at least it was used only in low-quality manufactures such as acid for manure works. In Devon and Cornwall, pyrites was derived from the cleaning of lead and copper ores. Wicklow pyrites came from very thick beds, particularly in the Vale of Avoca where one deposit contained as much as 44 per cent sulphur. Spanish pyrites was imported to the Tyne in 1856, and found favour there because although, as with home pyrites, the cost of roasting was about 2*s*. a ton, the sulphur content was 45–50 per cent compared with 35 per cent for Wicklow pyrites. By 1862 some 90 per cent of the sulphur used by the alkali industry in the country as a whole was derived from pyrites. (Gossage, 1862, p.105.) Five years later it was estimated that some 500,000 tons of sulphuric acid were made annually in the United Kingdom, 320,000 of it for the Leblanc process and most of the rest for the production of artificial fertilizer. For the production of this acid 160,000 tons of sulphur were required, but only 10,000–12,000 tons of it were brimstone, the rest being obtained from burning some 375,000 tons of pyrites. Although it required more space in the plant, more energy, and more labour, sulphur from pyrites by that time cost £4. 10*s*. a ton, compared with £6–£7 a ton for Sicilian sulphur. (Mactear, 1868, p.225; *C.N.*, 6 Nov. 1874, p.218; Table 8.)

Table 8. *Sources of pyrites imported into United Kingdom, 1873–82* (thousand tons)

	Norway	*Portugal*	*Spain*	*Other*
1873	67	119	247	7
1876	8	56	419	21
1879	8	82	374	16
1882	—	114	498	16

Sources: *Engineering & Mining Journal*, 18 Oct. 1884, p.264.

By 1867/8 half the pyrites used in Britain came from Spain, almost all of it from Huelva. In 1860 three foreign companies were working Huelva ores and the trade was already expected 'to change the whole face of a district, which had long been poor and destitute of resources' (*M.J.*, 24 Mar. 1860, p.197). The Tyne also used some pyrites from near Trondheim, and large amounts derived from the lead and zinc mines of Namur province and from near Spa in Liège province shipped from Antwerp at very low freight charges, sometimes as little as 6*d*. a ton. (Mactear, 1868.) Early in the 1860s total Tyneside consumption of pyrites was about 70,000 tons worth £105,000. (Campbell, 1962, p.29; *C.N.*, 17 Oct. 1863, pp.188–9; Tables 9 and 10.)

Table 9. *Main sources of pyrites supply to the Tyne, 1862/3* (thousand tons)

Huelva, Spain	30
Belgium	12
Westphalia	6
Hitteren Island, Trondheim	6
Wicklow	4
Coal measure pyrites	7

Source: J. Pattinson (in Armstrong, 1864).

Use of pyrites brought associated industries to the chemical districts. Most notable were the plants for copper recovery. Copper smelting itself had produced large tonnages of sulphur burned off at immense cost in labour and coal, and in environmental damage, particularly to Swansea and neighbourhood—in the early 1850s copper-smelting sulphur losses were put at 60,000–70,000 tons of sulphur a year (*M.J.*, 1 May 1852, p.212). By recovering sulphur, the alkali districts now first supplied the copper smelters with residue and then produced their own copper. Smelters were built on the Tyne, in south Lancashire, central Scotland, and at Oldbury. The remaining residue, rich in iron, was sold to blast furnaces. The establishment of the Steel Company of Scotland at Hallside in Lanarkshire was associated with a plan to use pyrites residue for steel-making.

Table 10. *British supplies of pyrites and brimstone, 1882* (tons)

Brimstone imports	47,278
Pyrites, home production	25,400
Pyrites imports	
Liverpool	261,085
London	49,912
Tyne district	131,952
Glasgow	25,520

Source: *J.S.C.I.*, 29 Sept. 1884, p.479.

Nitre

The other raw material used in sulphuric acid manufacture, nitre, came from saltpetre worked in the Chilean deserts and, before the war of 1879–82, in the desert areas of Peru and Bolivia. In the early 1880s as much as 100,000 tons of saltpetre were exported to Britain, largely through Valparaiso, from these 'Caliche' deposits. A quarter of a century later, nitrate of soda made up 70–75 per cent of Chile's exports. Long before this, however, the needs of the alkali trade had ceased to be a very significant factor in the demand for nitrate of soda. From 1859 onwards the Glover tower saved a considerable loss of sulphuric acid and reduced the significance of this already minor input.

Salt

Common salt was the key raw material in alkali manufacture. In its early development, essentially its prehistory, the very small mineral-based alkali trade used salt from brine springs. After the reduction of the Salt Duty in 1823, and particularly after its abolition in 1825, the Leblanc process depended on dry salt, usually rock-salt, but in some instances salt produced by the evaporation of brine. The Solvay process used increasing quantities of brine from 1874 onwards. In the early stages of development proximity to salt-fields and, later, the difficulty of moving large tonnages of brine in part determined the location of the alkali industry.

In the eighteenth century the coast of north-east England had been a centre of the manufacture of salt from sea-water, but this industry was in decline by

1800 and was never important for the alkali trade. In the nineteenth century the only nationally significant centres of salt production were inland. Seventy-five works produced fossil salt at this time: two were at Weston-on-Trent and Shirleywick on the Trent and the Grand Trunk Canal some four miles north-east of Stafford, six were in the Droitwich area of Worcestershire, and sixty-five in Cheshire, mainly in Northwich and Middlewich (Knight, 1851).

Rock-salt had first been proved at Marbury near Northwich in 1670 at a depth of 34 yards during a search for coal. At the end of the Napoleonic Wars Cheshire's annual capacity was put at 533,000 tons, but operations were then far below capacity, the average output of the five years to 1817 being only about 189,000 tons. There was much conflicting evidence on the wider economic impact of the industry, but one local manufacturer estimated at that time that 6,500 people were immediately dependent upon the salt trade and a further 5,000 worked on the flats and barges which carried salt to Liverpool and brought considerably in excess of 100,000 tons of Lancashire coal back to the salt-works and districts. (Bourne & Horne 1817.) By the thirties the Weaver was one of the most crowded channels of communication in Britain (McCulloch, 1839, pp.590–1). By the early 1850s there were about 600 boats in the trade, and at that time 12 cwt of coal were used to produce one ton of salt (*M.J.*, 10 Jan. 1852, p.21).

Weaver valley salt became the chief source of supply for the Leblanc alkali industry not only in Lancastria, but in all the chief producing districts in Britain. By 1881 production had reached 1,630,000 tons. Of this, 1.1 million tons were exported from the Mersey. Much of the rest was used in Lancashire alkali works or sent by canal across Britain for the Leblanc works on the Tyne.

Salt had been produced on only a small scale near Weston-on-Trent, and by 1881 output was as little as 4,000 tons. However, at this time Stafford Corporation, boring for water, proved new salt deposits, and in the following years a private concern developed brine pumping there. (*J.S.C.I.*, 31 Mar. 1891, p.281.) This became of some importance to the relatively small Black Country alkali trade. Brine production in the Droitwich and Stoke Prior districts of Worcestershire was on a bigger scale—215,000 tons in 1881—and in the second quarter of the nineteenth century had supported a notable local alkali trade. After 1850, however, production was confined to salt for sale.

The existence of salt in the Permian deposits at the mouth of the Tees was proved in 1862, but production was not begun there for another twenty-two years, the delay being largely due to difficulties of extraction. Salt in the Northwich area of Cheshire was first reached at a depth of 100–120 feet, but on the Tees it was 1,100–1,200 feet below the surface.

In 1881/2 remarkably pure rock-salt was discovered near Carrickfergus. The trade in salt from here was built up, but expectations of large deliveries to the soda-works of Scotland and the north-east of England were disappointed. (*M.J.*, 27 Nov. 1852, p.575; *Mineral Statistics of the United Kingdom*, 1855, p.96.) Hitherto the Irish linen trade had obtained its bleaching liquid from English or Scottish works; but hopes that home salt would stimulate a big home alkali industry were never realized. (*M.J.*, 15 May 1852, p.236.)

Salt was eventually proved in three other areas around the Irish Sea in association with small occurrences of Triassic rocks. At Preesall east of Fleetwood, the Fleetwood Salt Company started production in 1889. Rock-salt mining was begun in 1893. In 1907 half the output of mined salt was used in alkali works on the Mersey and Dee, and at Glasgow, Irvine, and Bristol. Much of the brine was employed at a Solvay alkali works built at Burn Naze west of the estuary of the River Wyre from the Preesall field. (Sherlock, 1921.)

Late in the nineteenth century a Liverpool syndicate, boring for coal near the Point of Ayre in the Isle of Man, discovered extensive salt beds. The Manx Salt and Alkali Company was formed to pipe the brine the six or seven miles to Ramsey. Commercial production began in 1903, but most of the salt was used in the fisheries, for butter production, or as agricultural salt. (Sherlock, 1921, pp.82–6; *C.T.J.*, 9 July 1898, p.23.)

By 1897 the Barrow Salt Company had begun to pump brine from the salt beds found below Walney Island. Initial plans were for a production of 50,000 tons of salt a year, and a large chemical works project was rumoured (*C.T.J.*, 2 Oct. 1897, p.215), but salt production never became very important and the chemical works were not built.

Throughout its history the Leblanc alkali trade in Britain remained dependent on bought salt. In 1852, for instance, there were 97 salt-works in the United Kingdom, but only one, in the Winsford area, was owned by an active alkali concern, Muspratts. (*M.J.*, 10 Jan. 1852, p.21.) When the United Alkali Company was formed in 1890, apart from wells on Teesside, it controlled no salt supplies.

Limestone

Limestone was used in large tonnages in the alkali industry. It was the essential flux in the black ash furnaces, and fine lime was used in the manufacture of bleaching powder. In Solvay's process the carbon dioxide for saturating the ammoniacal salt solution was derived from kilns charged with limestone and coke. The main supply came from the widespread Carboniferous Limestone outcrops. In north-west England the limestones of North Wales and more especially of the Peak District were the chief suppliers. Tyneside was unusual in making use of chalk shipped northwards as ballast from the Thames, and even in using chalk from France, but eventually it had to use limestones from the Pennines and the North of England.

CHAPTER 5

Alkali Economics and Location

Costs and prices

The nineteenth-century alkali industry operated in a setting of a rapid increase in output and falling prices. Changes in raw material costs were smaller and much less definitely downwards, and wages rose. The industry was therefore compelled to economize. Instead of making large profits from wide margins of prices over costs, it had to resort to larger out-turn on smaller margins (Fig.2 and Tables 11 and 12).

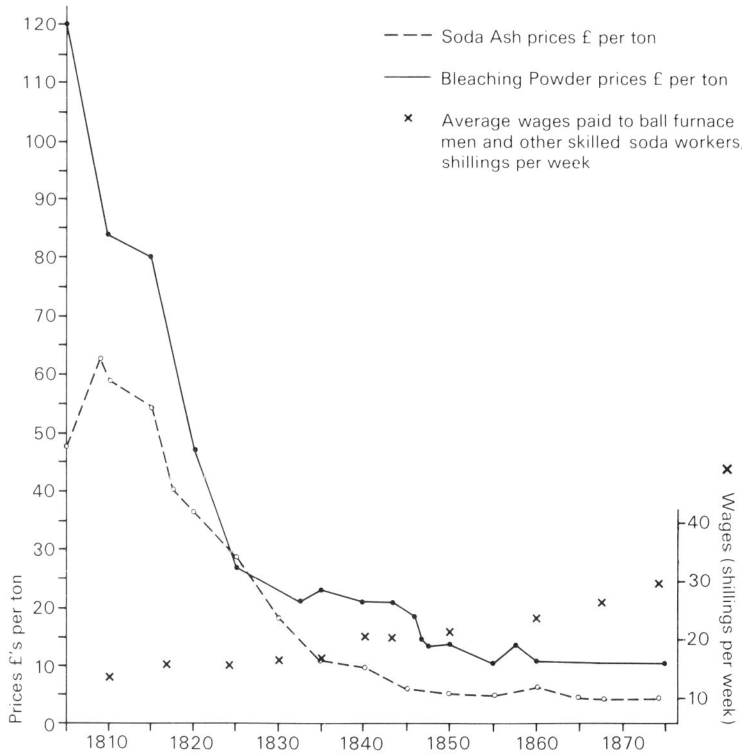

Fig. 2. Soda and bleaching powder prices and wages 1805–75 *Source: M.J.*, 6 Feb. 1875

Table 11. *Comparative prices and costs, 1827–86* (per ton)

	1827–30	*1839–40*	*1878*	*1886*
	£ *s. d.*	£ *s. d.*	£ *s. d.*	£ *s. d.*
Soda ash prices at Liverpool	16 10 0	15 0 0	5 10 0 to 6 0 0	4 0 0 to 4 10 0
	At Liverpool		*At Widnes*	
Material costs				
Sulphur	5 10 0	11 10 0*	6 15 0	5 0 0
Salt	6 0 to 8 0	9 0	8 9	8 0
Limestone	N.A.	4 6	5 6	4 6
Lime	13 4	7 6	14 0	11 6
Nitrate of soda	N.A.	16 10 0	14 0 0	9 10 0
Coal slack	6 0 to 6 9	5 0	5 0	4 6

*Period of Sicilian sulphur monopoly.
Source: Muspratt, 1886, p.410.

Table 12. *Estimated costs and returns in an alkali works making 200 tons finished product weekly, 1863*

Sales	*Amount (tons)*	*Price per ton (£. s. d.)*	*Total (£. s. d.)*
Soda ash	50	8 8 9	421 17 6
Soda crystals	90	4 15 0	427 10 0
Bleaching powder	30	9 0 0	270 0 0
Caustic soda	30	16 0 0	480 0 0
			1599 7 6
Sales costs			
Commission at 5 per cent			79 19 6
Discount at $2\frac{1}{2}$ per cent			39 19 6
Carriage of products at 15*s*. a ton			150 0 0
			269 19 3
Over-all revenue from sales			1329 8 3
Production costs			
Pyrites	162	1 2 0	178 4 0
Nitrate of soda	7	16 0 0	112 0 0
Salt	168	11 0	92 8 0
Coals	415	5 0	103 15 0
Manganese	32	3 2 6	100 0 0
Limestone for soda ash	190	4 4	41 3 4
Limestone for bleach	32	8 0	12 16 0
Casks			139 0 0
Men's wages			235 0 0
Expense and taxes	£650 per annum		12 10 0
Rent, salaries, etc.	£2730 per annum		52 10 0
Wear and tear and repairs	£1540 per annum		20 0 0
			1099 6 4
		Weekly profit	230 1 11

Source: *M.J.*, 21 Nov. 1863.

Processes were adjusted as prices fell. Immediately after 1800, soda crystals sold for £60 a ton; by 1863 they were £4. 50 to £4. 75 (Richardson & Watts, 1863, p.199). In 1824, when Muspratt first sold soda ash, the price was £25 a ton, and in 1861 only £8. The cost of the materials used had changed substantially in relative standing. Under the conditions of trade of the early Leblanc industry the gap between costs and selling prices was wide. In these circumstances even small works could live and thrive, but the hard years of the seventies and particularly the eighties narrowed the margins and quickly thinned the list of producers. However, not only did the balance of cost and price change over time, but at one and the same time various combinations of inputs were possible with consequent variations in over-all costs (Table 13).

Table 13. *Production costs for sulphuric acid at Tyneside alkali works c. 1860*

	£	*s.*	*d.*
(1) Using sulphur			
$5\frac{1}{2}$ tons sulphur at £7 per ton	38	10	0
5 cwt nitrate of soda at 18*s*.	4	19	0
2 men at 21*s*. per week	2	2	0
	45	11	0
(2) Using Belgian pyrites			
13.75 tons at 45*s*.	30	18	9
6 cwt nitrate of soda	5	18	9
8 men at 21*s*. per week	8	8	0
	45	5	6
(3) Using Irish pyrites			
18 tons at 33*s*.	29	14	0
6 cwt nitrate of soda at 18*s*.	5	18	9
9 men at 21*s*. per week	9	9	0
	45	1	9

Source: Richardson & Watts, 1863, pp.70–1.

The Leblanc industry did not require large scales of production, so that entry was relatively easy. By contrast the ammonia soda process involved larger scales as well as more sophisticated procedures and its capital costs of plant were considerably higher; as a result fewer plants were built. In 1880 there were 120 Leblanc alkali works and two ammonia soda works in Britain. Forty years later there were no complete surviving Leblanc works, five ammonia soda works, and one major electrolytic alkali plant. The 120 Leblanc works had been widely scattered throughout industrial Britain, though with a few major concentrations, the six surviving alkali works were within 55 miles of each other in Lancastria.

As late as the mid-1880s it was remarked of the Leblanc trade that 'the plant required is not of a very costly nature or character'. In the peak production year of 1880 the average value of British Leblanc works was only

£20,000. (*Industries*, 29 July 1887, pp.119–20.) Brunner Mond & Co. invested approximately £30,000 in 1873/4, and even in their second year of production made only 2,408 tons. Solvay himself reckoned in the 1870s that the capital cost of his plant was probably £1,600 for every ton of daily capacity or, with land included, of the order of £2,400. (Watts, 1923, pp.23, 31; Lunge, 1880, pp.32–3.)

Even within the Leblanc industry there were some scale economies. In 1880 it was suggested that the very smallest works should be designed for 130 tons of saltcake weekly. A small works was characterized as one which made up to 200 tons weekly. Five years earlier Chance of Oldbury was turning out some 900 tons a week, and Allhusens' works at Gateshead was even bigger. (Lomas, 1880, pp.4, 6; Chance & Hunt, 1875.) The bigger firms commanded more capital, and so could adopt technical improvements and meet competition that ruined smaller firms. In the middle years of the century, it was initially only the big producers who could afford to install plant to recover hydrochloric acid gas. Their outlay gave them a substantial process cost saving as well as reducing the nuisance problem and therefore the amount of litigation with which they were bothered. (Muspratt, 1886.) At a more critical time in the seventies and eighties, when depression coincided with the mounting competition from ammonia soda, it was again the bigger firms which mechanized and improved their operations to reduce costs and widen their product range. The big firms survived, the small units failed (Table 14).

Table 14. *Plant size characteristics in the Leblanc alkali industry illustrated by Number 1 District 1866, 1872, 1876, and 1882*

	1866	*1872*	*1876*	*1882*
Number of works	31	40	46	36
Tonnage of salt decomposed weekly	3282	6120	7117	7497
Average works salt consumpton (tons)	105.9	153.0	154.7	208.2
Median consumption (tons	90	112	132	185
Number of new works since preceding date	—	9	9	2
Number of works closed since preceding date	—	none	3	12
Number of existing works showing increased outputs	—	27	19	24
Number of existing works showing decreased outputs	—	1	12	7
Unchanged capacity	—	3	6	2
Average tonnage of largest five works	251.0	370.0	428.8	487.6
Average tonnage of largest ten works	205.1	302.5	335.4	413.2
Average tonnage of smallest five works	26.2	26.0	19.5	34.0
Average tonnage of smallest ten works	35.1	40.5	36.8	55.0

Source: Annual Returns of Alkali Inspectors, *B.P.P.*

Integration backward to control raw materials was, as noted earlier, rare in the Leblanc alkali trade. Muspratts, Tennants, and the Netham works at Bristol were exceptional in controlling some of their mineral supplies. On its formation in 1890, the United Alkali Company had to purchase all its salt for the Lancashire operations, though it very quickly bought the Fleetwood salt-

1. The Mechanisation of an old technology – a revolving black ash furnace at Widnes Alkali Works (courtesy of I.C.I.).

2. The Landscape Impact of the Leblanc Process. St. Rollox Works, *circa* 1880 (from Cassell's *Great Industries of Great Britain 1882*).

3. The Castner–Kellner Electrolytic Alkali Works, Weston Point, Cheshire.
a. The greenfield site by the new Manchester Ship Canal.

b. The new works in the late 1890's (courtesy of I.C.I.).

field. U.A.C. was always dependent upon bought coal, even refusing to buy north-eastern collieries offered to it in the 1890s. Even though the ammonia soda industry came eventually to control limestone supplies in the Peak District, it too bought fuel. In its lack of backward integration, the alkali industry stood in marked contrast to the iron and steel trade.

Site conditions

All Leblanc works required land for the large-scale dumping of waste. In 1880 Lunge put the minimum area at 5 acres, but a few years earlier the St. Rollox works occupied 50 acres and had 'deposit land' and other land beyond the works built-up area amounting to 80 acres more (Lunge, 1880, p.325). It was important to have the facility for running away weak acid or alkali liquor. Rail or, better still, water access was also important. As early as 1836 in a public notice for sale of Heworth works on the Tyne it was considered worth stating that it was a quarter of a mile from a quay with eight feet of water and quarter of a mile from the Brandling Junction Railway then being built (*Newcastle Courant*, 23 July 1836). The quay at Allhusens' works, a little way above Heworth, was later capable of handling vessels of 300 tons burthen (Fordyce, 1857, vol. 2, p.784). The need for dumping land and the emphasis on access to bulk transport facilities in the choice of site were significant reminders that the alkali trade was a bulk-reduction industry and consequently one for which locational choice was of great importance.

Location

'The cheapness with which alkali can be manufactured depends on the locality. Four elements enter into the composition of alkali—salt, pyrites, coal and limestone and chalk. There are very few localities in which these several products are found in close proximity.' With these words H. Vivian summarized the locational problems of the alkali business to the House of Commons at the time of the debate on the Alkali Bill in 1863 (*M.J.*, 18 July 1863, p.514). Unfortunately he ignored the location of the market, and gave a static not a dynamic view of locational considerations (Fig. 3).

Reviewing the economic history of nineteenth-century chemicals in Western Europe, Hohenberg pointed out that local raw materials were important in localizing early developments of manufacturing, as with the soda industry near the salt and limestone of the Zurich district or near numerous small coal deposits in south-eastern France. With greater scales of operation, emphasis was placed on efficiency of location for cheap bulk transport from bigger or richer material sources. In this way, Basle replaced Zurich, and Stuttgart was replaced by Mannheim (Hohenberg, 1967, p.50). A factor which Hohenberg might well have added was the supersession of local and regional markets by national ones. Thus the growth of the major alkali industry of the second quarter of the nineteenth century in the Marseilles district to supersede Paris, was a response not only to local limestone, salt, and small coal deposits, but to

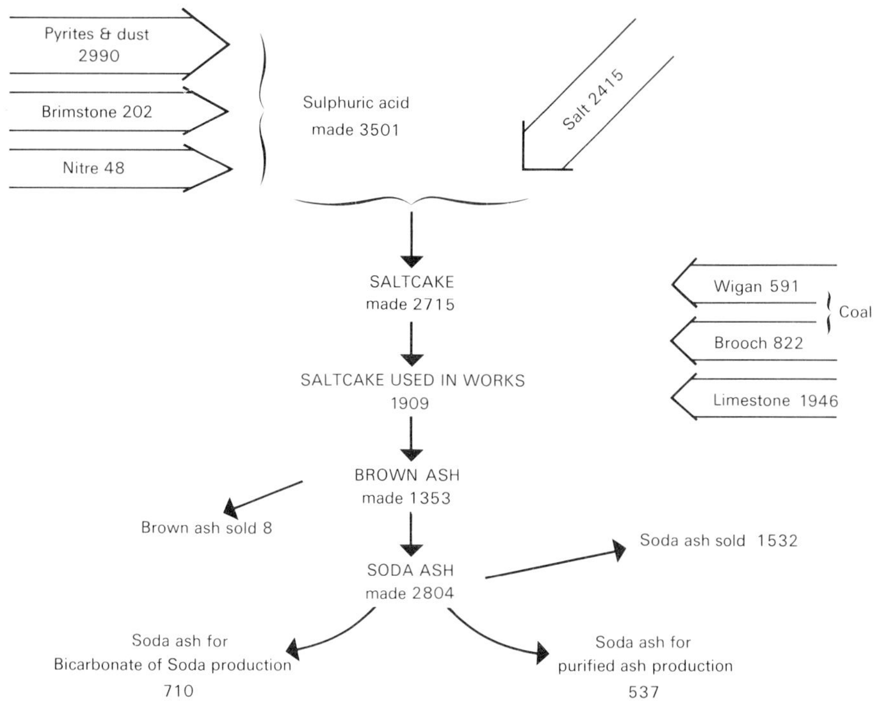

Fig. 3. Chances of Oldbury. Flow chart of materials and products 1875 (tons) *Source:* Chance & Hunt, 1875

the major local demand for alkali in soap-works. Zurich town and canton, though superseded later by Basle, formed a great textile centre, silk and cotton employing many thousand in the 1840s. In the case of British alkali, international markets were also of great importance.

The main British alkali districts—Tyneside, Merseyside, and Clydeside—were well placed for local mineral supplies, and in the early development of the industry, when scales of production were still small, had important demand from local consuming trades, notably glass- and soap-making and textile manufacture. However, they were also well located to cope adequately with long-distance, large-volume movements of raw materials, by coasting trade to serve other British markets, and, above all, for access to world markets. Other areas possessed some of these advantages, but not all of them, and, after promising starts, either faded out altogether or played a limited role (Fig. 4).

Time is important in location in at least three aspects: in the evolution of material balances within the industry, often as a result of a steady increment of small-scale changes; in the introduction and diffusion of radical new techniques; and in the stagnation which seems to have affected old firms, plants, or even districts.

Fig. 4. Leblanc alkali works in Britain 1865–9 *Source:* Reports of Alkali Inspectors

In material inputs, the most important change, and one which had major locational significance, was a very marked fuel economy. Partly because of its non-continuous nature, partly because of the character of the plant and processes involved, Leblanc alkali manufacture was fuel intensive, indeed profligate. Isaac Lowthian Bell (1871, p.222) revealed something of its prodigality and of the possibilities for saving energy to the 1871 Coal Commission. He described the small and inefficient saltcake furnaces which had wasted fuel in the past and observed how, in order to economize in energy, flue temperatures had already been reduced from the old level of 1200–1500°F to 700 or 800°F. The reduction of fuel consumption over a longer period was most pronounced (Table 15).

Table 15 *Consumption of materials for one ton of soda ash in Lancashire works, 1861 and 1886* (cwt)

	1861	*1886*
Pyrites	24	13.5
Saltpetre	1	0.21
Salt	25	25
Limestone	30	26
Coal of slack	70	55
Total	150	119.7

Coal remained the largest material input, though collectively other raw materials bulked even larger. Clearly an ideal location was on or near a coalfield, but with good access, preferably by water, for other mineral raw materials, and, as scales of production grew, with good access also to major home markets and to ports which had efficient services to the markets in the major industrial nations overseas. As the Leblanc trade passed its apogee and was beginning to decline, its locational needs were summed up by Lomas (1880):

> A moment's thought will suffice to show that it is hardly possible to be in the immediate neighbourhood of the staple raw materials and of the market for the finished goods. Of the two the former is to be chosen, because the conveyance of pyrites, coal, salt, limestone and other bulky articles from a distance is very much more expensive than that of the final products—soda ash, crystals, caustic soda, bleaching powder etc.—which are smaller in amount, and more compact in package. Other matters being equal, and an easy conveyance of the outlying raw materials being at command, *proximity to a good coalfield* is the first thing to be desired.

Technical advances such as the great economy in the use of nitre by means of Glover's tower, or the even greater saving involved in Weldon's process of chlorine production, helped innovative firms and sometimes had a significant, though not always predictable, effect on locational values (Findlay, 1965, p.85; *M.J.*, 13 Feb. 1875, p.162). The development of a practicable ammonia soda process brought a decisive change. Brunner and Mond both had long experience of Leblanc soda production at Widnes. In 1882 Mond compared the inputs and the costs of production for Solvay soda ash at a theoretical plant at Winsford with those for a Leblanc plant at Widnes (Table 16).

Table 16 *Material inputs for the Leblanc and Solvay processes, 1872* (cwt per ton of soda ash)

	Leblanc	*Solvay*
Salt	29	40 (as brine)
Limestone	35	44
Coal slack and coke	88	44.5
Pyrites	17.5	—
Nitrate of soda	0.5	—
Ammonia sulphate	—	1.80
Total	170.0	130.3

Source: Cohen, 1956.

Apart from some completely new material needs—sulphate of ammonia for ammonia production in place of pyrites, for instance—there were two major changes. One was the substantial increase in salt consumption, and, decisive in locational terms, its requirement in the form of brine rather than of salt. The other was the reduction in the consumption of fuel by perhaps 50 per cent. In short, the new techniques emphasized the value of a local brine-field rather than a coalfield or even an intermediate location. Suggestions made both on Teesside and in Cheshire that brine should be piped to the older alkali districts were firmly rejected by local interests eager to promote the development of the mineral district (Calvert, 1915, p.ix). Later progress led to very pronounced fuel economy and also a reduction in the consumption of limestone and this increased the advantages of the brine-fields. By World War I Partington (1919, pp.85ff.) reckoned that, in addition to brine, each ton of Solvay soda needed inputs of only 5 lb of ammonia, 1.3 cwt of coke, 10 cwt of coal, and 24 cwt of limestone. The electrolytic processes introduced yet other considerations. Low-cost power sites were ideal, but in view of established patterns of ancillary trades, markets, the remoteness of hydro-power sites, and, perhaps, the prevailing cautious entrepreneurship of the 1890s, it is not surprising that the British electrolytic industry moved to the old alkali districts (Fig. 5).

Locational choice is rarely if ever wholly logical. The entrepreneur cannot know all the facts about the cost situation when his planned plant will come into production, or about the locational decisions which his rivals are taking. Moreover, limited perception on the part of the entrepreneur was, even in Victorian days, matched by a less than consistently optimizing motivation. Mond (1869, p.235) estimated that, for a capital outlay of £80,000 on sulphur recovery, alkali firms could make a clear annual profit of £160,000, but few were interested. Knowledge was limited, rationality bounded, and the aim satisfactory rather than maximum profit. Locations were chosen which a strict locational analysis would have shown to be less than ideal. Conversely, works once established were not readily given up, although trends could be very diverse.

Often, as conditions worsened, firms accepted lower profits or even losses—

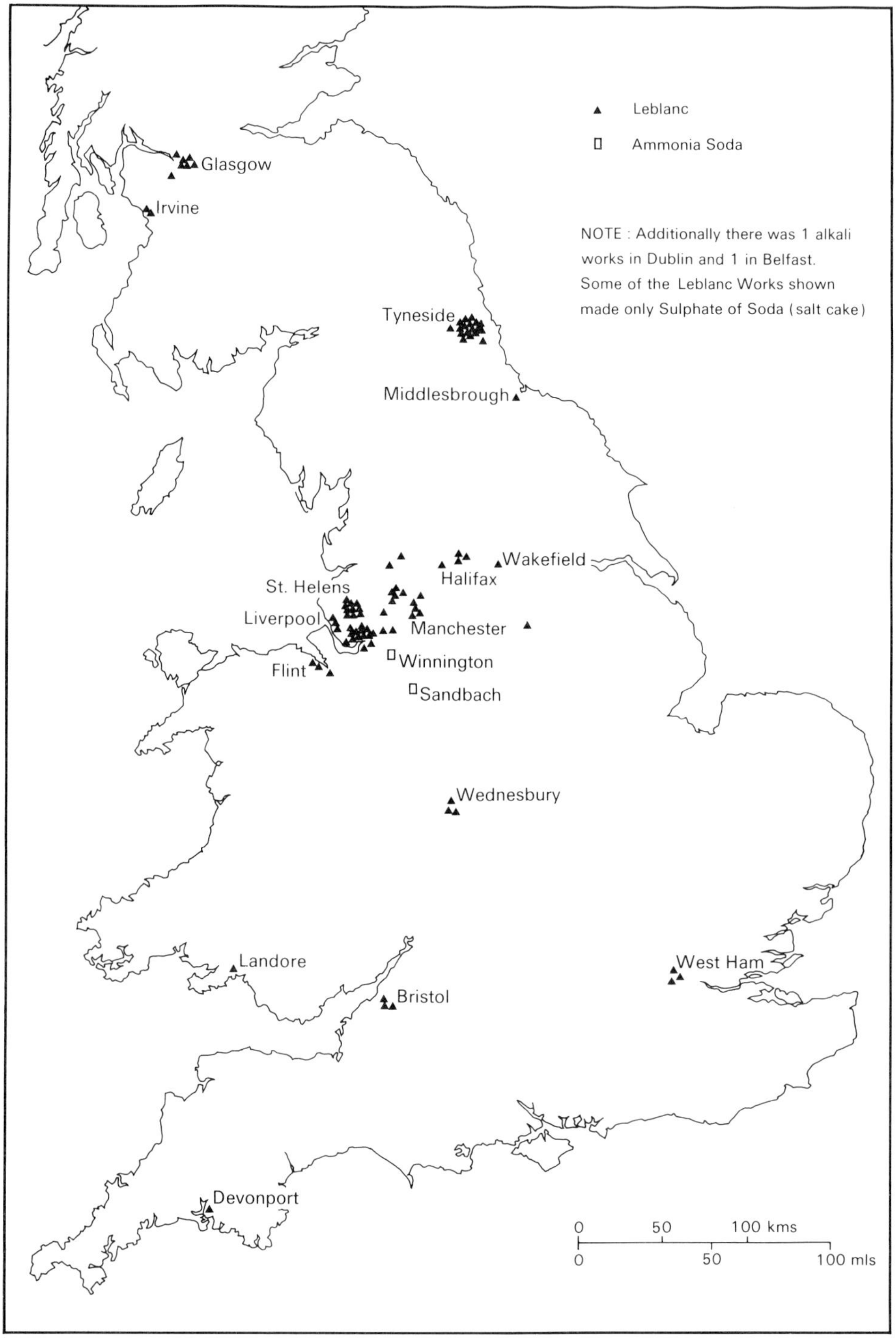

Fig. 5. Alkali works in Britain 1882 *Source:* Alkali Statistics (1882), *B.P.P.*

in 1897, for instance, it was shown that saltcake manufacture involved a loss of 12*s*. a ton (*C.T.J.*, 1 Jan. 1898, p.1). Other works reacted positively in pioneering technical innovation to counteract the disadvantages of their locations. Overall there was a high mortality of firms. In 1878 David Gamble of St. Helens reckoned that half those who had entered the alkali trade in Lancashire had failed (Gamble, 1878, p.206). Apart from a willingness to accept lower returns (Mond's cost estimates suggested that in 1872 when soda ash was selling in Liverpool for £10.50 a ton, Widnes Leblanc costs of production and delivery to Liverpool were £9.35, but Solvay works should be able to deliver it for £6.94), there was a good deal of manœuvring, ways in which overall costs could be reduced by action in the field of process costs. Location affected but did not determine firm survival and degree of success.

CHAPTER 6

Minor Centres of the Trade and the Scottish Industry

In 1852 John Williamson of the Jarrow Chemical Company attempted an estimate of the state of the alkali industry in Britain, though on the basis of incomplete returns. He reckoned that 137,000 tons of salt were decomposed by chemical works annually. Of this, 71 per cent was consumed in the works of either north-east England or Lancashire and a little under 14 per cent in west Scotland; the remaining 15 per cent was divided between a scatter of operations. Merseyside, though not pre-eminent until after mid-century, can be seen to have had the best combination of factors. Tyneside, leader until the fifties, lacked a regional supply of salt until it was already in decline, and the local soap, glass, and paper trades shrank and were eventually either extinguished or dwarfed by those of Lancastria. Clydeside enjoyed wide trade links, but had no salt, and chemical manufacturers faced keen competition for fuel supplies from the Lanarkshire metal industries. In Scotland there were relatively few home outlets for alkali. All the other districts had even fewer advantages for the trade.

Most of the firms in the minor centres were established to supply local demands. Commonly they were small producers. By 1881/2 when an official list of chemical works of all kinds was published, there were isolated alkali works in the South-West (Devonport) and in South Wales (Landore) and three in East London, in the old heavy industrial core area around the outfall of the River Lea into the Thames in West Ham, Poplar, and Silvertown. The other two and bigger market-oriented agglomerations were in the West Riding and in the textile districts of Lancashire well away from the mid-Mersey chemical centres (though unfortunately included in the county totals for production).

The bleaching and dyeing activities of the woollen and worsted districts of west Yorkshire, its important glass industry, and its own abundant coal supplies favoured a local alkali industry. Leeds played a not insignificant part in the development of the trade's technology. It was a puzzle to the mid-nineteenth-century local enthusiasts that no larger chemical industry emerged. The explanation is perhaps partly to be found in the cumulative effect of regional specialization of manufacturing, the West Riding was a textile, quality iron and steel and engineering district, there was no capital or expectation for a major chemical activity as well. By 1882 Yorkshire had five alkali works, one in Rotherham and four in the textile district, all of them reasonably centrally placed within it, one at Alverthorpe 2 miles north-west of Wakefield, one in Birstall, and two in Cleckheaton.

East Lancashire was more important. Here there were thirteen works altogether by 1882. Four were in the heavy industrial eastern edge of Manchester at Clayton, Ardwick, Miles Platting, and Gorton Brook. Some proved to have good survival properties. In the 1830s Tennants of Glasgow bought the Ardwick Bridge Chemical Works. Howarth and Company of Miles Platting, established in 1854, by the early 1890s were especially noted for saltcake produced for glassworks. Of the other nine Lancashire works two were in the weaving towns of the Calder valley, both near Accrington. In 1882 seven were south of Rossendale, at Chadderton, Westleigh between Hindley and Leigh, and the remaining five interestingly concentrated all within some two miles of Farnworth station south of Bolton. The reasons for the Farnworth concentration appear to lie in the textile history of the district. The Bolton area had some of the largest bleaching works in Britain. To serve them a number of sulphuric acid works had already been built near the town and several more later became Leblanc alkali manufacturers. (Lewis 1831; Hardie & Pratt, 1966, p.26.)

Other Lancashire works evolved in close association with the dominant textile trades of the neighbourhood, gradually supplying a wider range of their chemical requirements. Thus the Richard Bealey works at Radcliffe near Bury were derived from a bleaching operation dating back to 1750. In 1793 Bealeys began to make sulphuric acid for bleaching purposes. In the nineteenth century they extended to the whole Leblanc process and by the early 1860s were selling twenty to thirty times as much chemical as they used (Noxious Vapours, 1862, p.113). Later, when pressed to join the United Alkali Company, Bealeys refused and continued to expand in their existing lines of business, adding a caustic soda plant in 1895. Rileys of Hapton, on the Leeds and Liverpool Canal between Accrington and Burnley, provided an interesting illustration of the progression from the textile to the chemical activity. The Riley family were engaged in the dyeing trade of Rossendale. In 1834 they went into calico printing at Accrington, but nine years later gave this up to produce nitric and hydrochloric acid and dyers' solutions—for paper-makers, bleachers, calico printers, and dyers—at the little village of Hapton. They installed sulphuric acid chambers in 1848, and three years later first produced refined soda ash, which they delivered to almost all the calico printers and dyers in the area. Later they made caustic soda. By 1900 Rileys were still making soda ash and caustic soda by the Leblanc process (*C.T.J.*, 20 Jan. 1900, p.51).

The West Midlands and Severn Valley centres of alkali production

There were three important focuses of alkali manufacture in the belt from the West Midlands conurbation to Bristol. At Droitwich and Stoke Prior the industry made an impressive early start but then disappeared completely in the middle of the nineteenth century. Salt had been produced in Droitwich for many centuries. In 1828 it was proved near Stoke Prior some $3\frac{1}{2}$ miles to the north-east. Initially mined, it was later worked as brine. Stoke Prior had the substantial advantage over Droitwich of being on the line of the Birmingham and Worcester Canal, and so was able to obtain deliveries of coal by barge. In 1830 William Gossage, a Chesterfield-born chemist, became associated with Jonathan Fardon in the manufacture of salt, vitriol, soda, and soap from the Stoke Prior brine. Here in 1836 Gossage developed his condensing tower for hydrochloric acid gas. (*C.N.*, 4 Apr. 1877, p.186; I.C.I. Stoke.) By the mid-1840s Gossage and Fardon's British Alkali works occupied nine acres of canal-side land and represented the very considerable investment of £200,000. (Fardon, 1845, pp.169, 366; Lewis, 1848.) On the other side of the canal the Imperial Alkali Company was formed in 1836.

In the 1840s difficulties began to afflict the Stoke Prior alkali and salt industries. Located in a rural area, they had no local markets, and therefore long hauls—there were 1,200 tons of weekly traffic in and out of the British Alkali works, including 500 tons of coal, mainly from Brierley Hill but some from as far away at Staveley near Chesterfield (Fardon, 1845). In the 1840s water began to seep into the brine and reduce its density. Gossage meanwhile was diversifying his interests, making white lead in Birmingham in 1841 and smelting copper in South Wales a little later. He returned to Stoke Prior only in 1848. (*C.N.*, 4 May 1877, p.186.) By the late 1840s the Stoke Prior firms were collapsing. The Imperial Alkali Company's affairs came before the Court of Chancery in 1850. It was shown that the company had over-extended itself in the soap business and had been in debt after 1840. In its history of fourteen years no dividend had been declared. The works was sold by auction in 1851. (*M.J.*, 2 Nov. 1850, p.519; I.C.I. Stoke.) In 1850 Gossage moved to Widnes to set up alkali works there. After this, with John Corbett, later known as the 'Salt King', as the controlling power, Stoke Prior concentrated on salt. Its role in the alkali business had effectively ended by the middle of the nineteenth century. By nurturing Gossage's ambitions and abilities Stoke Prior had earned a place in the records of the British alkali industry.

In Bristol alkali production proved much more tenacious. Two of the twenty-three sulphuric acid chambers in England were in Bristol in 1820. At what was then called Natham Dam near the junction of the Feeder Canal and the River Avon, Netham works was established in 1847, one of its first three directors being a soap-maker. Within fifteen years its gaseous waste had become a great nuisance in Brislington parish. By the 1860s it had a wide product range, decomposing 100 tons of salt weekly in 1862 and 140 tons by 1867—at which time Bristol had three other very small Leblanc works. By now its own coal-mine had ceased to supply it with fuel, but from 1862 for sixty years the works controlled limestone quarries in the gorge near the Clifton suspension bridge; thereafter it obtained limestone from Cheddar. Netham delivered bleaching powder to local paper-mills and hydrochloric acid to Bristol galvanizers. In 1882 two other Leblanc works were still operating in Bristol, though their product range was more limited than that of Netham, ending with sulphate of soda.

After Netham joined the United Alkali Company in 1890, it served the West of England markets of that firm. Its activity held up well to a late date, even though from 1904 onwards it made no soda ash or caustic soda and concentrated on crystals made from Fleetwood soda ash. Netham was closed in 1949. (I.C.I. Netham; Polack (n.d.); *Bristol Times and Mirror*, 22 May 1923; Table 17.)

Table 17. *Netham Works, main products, 1910, 1925, 1940* (thousand tons)

	1910	*1925*	*1940*
Sulphuric acid	20	5	5
Saltcake	17	13	5
Soda crystals	13	10	7
Bleaching powder	3	2	—

Source: I.C.I., London Records.

Black Country alkali

There were three Black Country alkali works by the early 1880s of which the most important were those of William Hunt of Wednesbury and Chance Brothers of Oldbury. The Chance family acquired the Spon Lane glassworks in Smethwick in 1825, and produced alkali there in the early 1830s. However, having no land available for adequate expansion, in the mid-1830s they built a plant at Oldbury, in the hope of producing their requirements of sulphate of soda. At Oldbury they first burned salt with copperas (ferrous sulphate) but, the results being unsatisfactory, they quickly substituted the Leblanc process (Fig.6). Much trouble was caused locally by damage to gardens and crops but, as elsewhere, when the possibility of moving the works to another area was mooted, the local population became more amenable. Oldbury could obtain local Silurian limestone, although later more distant suppliers were probably the main ones; in 1906, for instance, Chance had a Great Western Railway quotation for delivery from Llanfynydd near Caergwrle in Flintshire. Coal was also available in the vicinity, though some was brought from distant sources—in 1853, for example, Staveley slack was being used and, in the first quarter of 1875, both Brooch coal and Wigan coals. In 1854/5 the firm was negotiating a rate for Cornish pyrites delivered by sea to Gloucester and from there presumably by canal; in the nineties it seems sometimes to have got its sulphur via the Great Western Railway from Cardiff, Penarth, and Barry. (Chance & Hunt, Carriage.) Salt supply presented difficulties. Droitwich was not far away and was linked with the Black Country by canal, but after 1888, when the Salt Union took over there, salt prices rose. Even though saltcake manufacture was continued, Chance and Hunt in 1899 also had quotations for delivery of saltcake from Bowman Thompson of Lostock Gralam at 7*s*. 11*d*. a ton in boatloads (Chance & Hunt, Carriage). A reliable salt supply was at last secured in 1907 by purchase of salt-works at Stafford.

William Hunt worked for Fardon and Gossage at Stoke Prior. In 1851, after the collapse of the Stoke Prior industry, he moved to the Black Country and

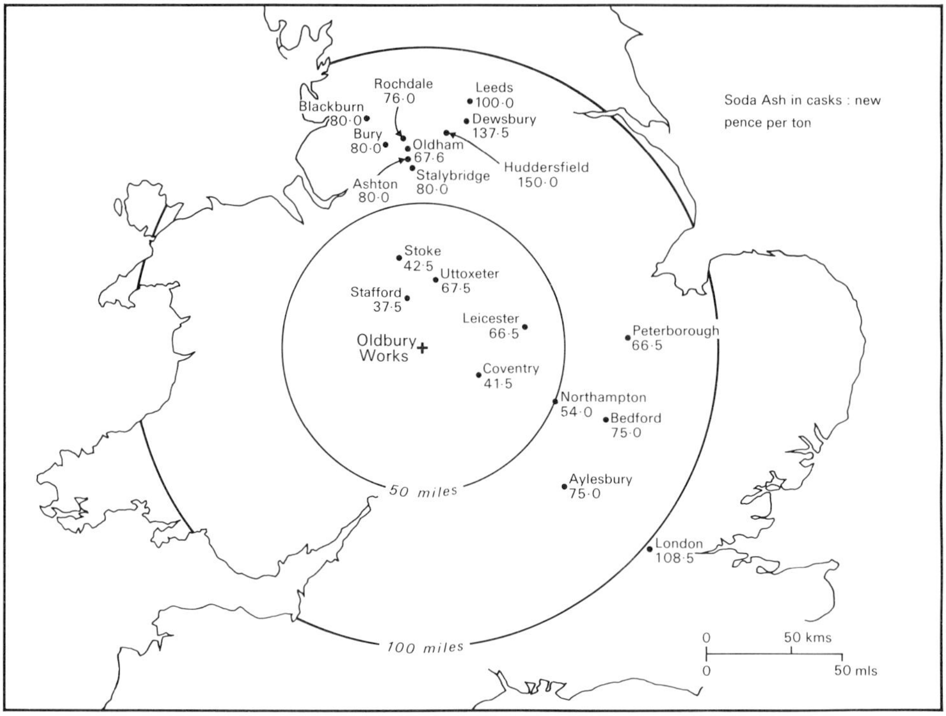

Fig. 6. Rail freight rates from Oldbury Chemical Works 1850 *Source:* Chance & Hunt, Carriage

built a small works to produce Leblanc alkali, soap powder, and sulphuric acid at Leabrook on the edge of Wednesbury. Later William Hunt put up a chemical works at the important coal and glass bottle town of Castleford on the Aire in the West Riding, but his family retained Leabrook. In 1890 Chances became the Oldbury Alkali Company, and was then the most important Leblanc firm to stay outside the United Alkali Company. Eight years later an amalgamation with the Wednesbury firm produced Chance and Hunt. Chance and Hunt did well as one of the few independent alkali firms, and grew by acquisition. A controlling interest in the Black Country firm was acquired by Brunner Mond & Co. in 1917. Together Chance and Hunt produced soda ash from 1835 to 1912, caustic soda from 1848 to 1919 from their own saltcake and, until 1923, from bought soda ash. Soda crystal manufacture ceased in 1912 after sixty years of production. (Chance & Hunt, Statistics.)

Scotland

Scottish works, almost all of them within the Clydeside industrial region, played an illustrious part in the development of the alkali industry (Fig.7).

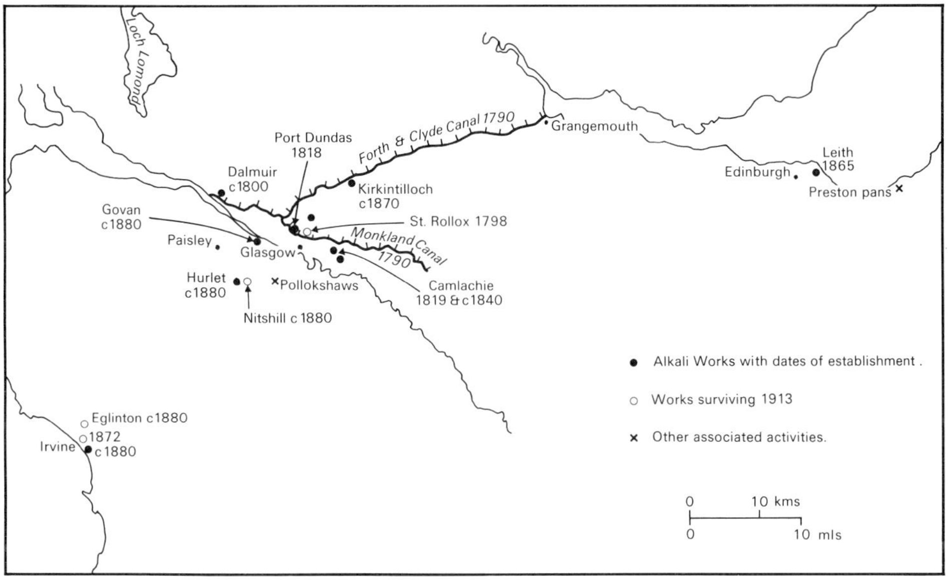

Fig. 7. Scottish alkali works during the nineteenth century *Sources:* various

The industry here illustrates the strength of change and also the very strong influence of continuity in chemicals. In the mid-1820s St. Rollox was the biggest works in the country with almost one-third of national output. Estimates made in 1852 suggested that Clydeside decomposed half as much salt as Lancashire and about a third as much as the Tyne. (Richardson & Watts, 1863, p.318.) Yet by the mid-1860s St. Rollox works was only half the size of Hutchinson's Widnes works, according to Mond. (Hardie, 1950, p.95.) Thirty years later Scotland was just a minor factor even within the rapidly shrinking Leblanc alkali industry, and it had no share at all in the Solvay process (Fig.8).

In the late eighteenth century Scotland was a promising field for the growth of the alkali trade. Although it lacked rock or brine salt deposit, it had a widespread sea-salt industry (Mactear, 1877). The first sulphuric acid was made in Scotland by the vitriol works of Roebuck and Garbett erected in 1749 at Prestonpans, long a great centre of the salt trade (Butt, 1967, p.254). By the 1840s, in addition to its own sea-salt industry, Prestonpans was importing rock-salt on a considerable scale from Liverpool and combined soap-making with its acid and salt activities (Lewis, 1847). In Glasgow there were already six to eight works making acid by 1797 and other important centres on the Forth as, for instance, at Burntisland (Mactear, 1877, pp.14–15). Coal was widely available and new communications opened more distant sources of supply. Between 1768 and 1790 the Forth and Clyde Canal was opened from Grangemouth to Bowling. In the latter year the Monkland Canal was com-

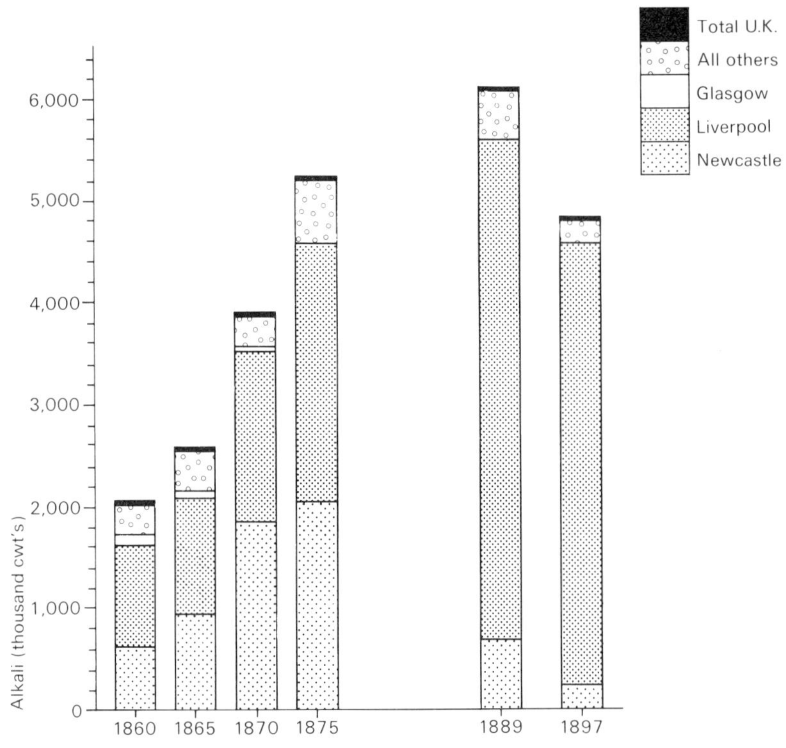

Fig. 8. Exports of alkali by principal ports 1860, 1865, 1870, 1875, 1889, 1897 *Source:* Annual Trade Returns

pleted from the heart of the Lanarkshire coalfield to meet a short branch of the Forth and Clyde at Port Dundas in North Glasgow. The Western Isles were one of the main sources of kelp and West Central Scotland had important alkali-using trades, notably soap and textiles. Cookson of Newcastle had introduced flint glass production to Glasgow in about 1777. The first growth points for the alkali industry were water-transport-oriented both in Glasgow and on the banks of the Forth and Clyde Canal near its outfall into the Clyde.

By 1800 Lord Dundas had a works at Dalmuir Shore in Old Kirkpatrick parish, built under the superintendence of the Earl of Dundonald, a significant innovator in the early British alkali trade. Dalmuir used soap-makers' lees from Paisley and elsewhere, formerly run off as waste, as its source of alkali. By the mid-1840s it employed 80–100 men and made 30 tons of sulphuric acid weekly which was used to decompose salt and to produce as finished products bleaching powder and soda crystals. By then it had become a blot on the Clydeside landscape, 'its furnaces and chimneys contrasting remarkably with the surrounding scenery which is very pleasing' (Mactear, 1877, p.24; Lewis, 1847; and Dumbarton, 1845). The direct descendant of

Dalmuir seems to have been the North British Chemical works with a Clydebank address, which turned out vitriol, caustic soda, chlorate of potash, and other products in the early 1880s. It was taken into the United Alkali Company in 1890 and closed within two years.

In Glasgow alkali was made as early as 1798, at first by 'furnacing' kelp with charcoal and a little quicklime (Mactear, 1877, p.23). Large-scale development began only with the enterprise of Charles Tennant. In 1787 a chlorine bleaching works had been built in Aberdeen, only two years after Berthollet had discovered the bleaching properties of the gas. Tennant, who had been apprenticed as a weaver, and later managed a bleach field at Darnley south-west of Pollockshaws, discovered and patented a process for using the bleaching properties of chlorine by moderating its action through the use of lime (and thus producing bleaching powder) which was much more convenient than Berthollet's method. In 1798 he moved to a new site on the Monkland Canal at St. Rollox, and there during the next year built the bleaching powder works which was to become the biggest chemical works in Scotland and to survive for over 150 years. Four other alkali works were put up in Glasgow—all of them small (Butt, 1967, p.138). Later six other works were built in the Clyde district, including plants at Kirkintilloch, Springburn, Govan, and two south-east of Paisley (Ferguson, 1876; Alkali Statistics, 1882). Most had short lives.

The most important focuses of the alkali industry outside Glasgow were Irvine and Eglinton on the margins of the Firth of Clyde. Irvine became an attractive overflow for Glasgow manufacture with the opening of the Glasgow, Paisley, Kilmarnock, and Ayr Railway which gave it excellent access to the business capital. Kelp, salt, and soda had already been made on the shores of Ayrshire, and by the mid-forties Irvine had 'some works for magnesia and other chemical processes' (Lewis, 1847; National, 1847). By 1876 it had two of the five active alkali works in the Glasgow district (Mactear, 1877). In 1872 W. Henderson had built the Irvine Chemical works, which by the mid-1870s was a substantial operator handling 20,000 tons of raw materials a year—one-quarter as much as St. Rollox—6,000 tons of coal (Ferguson, 1876, p.224). Alexander Walker, who came to manage Henderson's works, later set up the Irvine Chemical Company works and produced bleaching powder and caustic soda (Alkali Statistics, 1882; *C.T.J.*, 21 Nov. 1891, p.321, 15 Aug. 1896, p.101.) Eglinton seems to have been built in the 1880s.

There were undoubtedly many advantages from being in Irvine rather than in East Glasgow: waste disposal was easier; access to some raw materials was direct. At the beginning of the eighties the harbour, long silting, could only handle vessels of 100 tons burden, but in the next ten years it was improved. Chemicals were directly exported. (Lewis, 1847; Chamber, 1881; Popular, 1892.) On the other hand, although fuel prices were low, wages seem to have been above the Glasgow level and material inputs also on the high side. In the late 1880s fuel consumption per ton of saltcake charged to the black ash and balling furnaces was commonly 8–14 cwt; at Irvine works it was 13 cwt. (Dittmar & Paton, 1887b; Walker (n.d.); St. Rollox, 1893.) Levels were high in other processes too (Table 18).

Table 18. *Material inputs per ton of sulphuric acid at St. Rollox, Eglinton, and Irvine works, Jan.–June 1893* (cwt)

	St. Rollox	*Eglinton*	*Irvine*
Pyrites	9.80	10.75	11.69
Nitrate	0.18	0.17	0.29
Fuel	3.12	3.86	4.83

Source: St. Rollox, 1893.

In the half-year to June 1896 the packed cost of caustic soda at thirteen United Alkali Company works ranged from a low of £4. 12*s*. 11*d*. to a high of £6. 18*s*. 10*d*.; at Irvine the cost was £5. 12*s*. 9*d*. Profit per ton at £1. 10*s*. 10*d*. ranked just half-way in the list. In bleaching powder profit per ton was 17*s*. 11*d*. compared with 19*s*. 11*d*. at St. Rollox (U.A.C., 1896). Irvine works was closed in 1906, Eglinton in 1914, and Hendersons in 1920. (Ayrshire, 1951, p.105.) St. Rollox was both more successful and much more important.

St. Rollox works

St. Rollox works, as Mactear recalled eighty years later, was in its first years 'quite in the country', until Glasgow spread round it. Tennants began with and was for long most distinguished for bleaching powder. Output rose from 52 tons in 1799/1800, the first year of production, to almost 1,000 tons in 1825 and over 9,000 tons by 1870. Until 1803 Tennants bought their vitriol, partly from local makers, partly from Prestonpans, and even from as far away as Halifax; in that year six lead chambers were built, and by 1811 there were thirty-two. From 1803 a modified—probably much modified—form of the Leblanc process was operated at the works, but the full process was not used until 1818. Soda crystal production began in the latter year when 100 tons were made; by 1829 output was 1,400 tons. In 1825 soda ash was made; caustic soda followed in 1844, but demand proved inadequate and production was given up, though revived in 1875. (Mactear, 1877, pp.23–6; Tennant, 1922; Butt, 1967; Cassell, 1882, I, pp.296ff.) In the mid-nineteenth century the rate of expansion at St. Rollox was rapid; coal consumption, for instance, increased from about 30,000 tons in the mid-1840s to more than 125,000 tons some thirty to thirty-five years later. (Lanark, 1845; Noxious Vapours, 1878, p.275; Table 19.)

Tennants' works grew to such a size that it fascinated many observers. To the author of the Glasgow section of the *New Statistical Account* in 1845 the 'immense chemical works' were conducted on a scale of 'magnitude and perfection by Mr. Tennant', and it was reported that they were the most extensive of their kind in Europe. A decade later their buildings were 'black and dirty;

Table 19. *Materials and production at St. Rollox works, 1871/2* (thousand tons)

Inputs	
Coal	75
Salt	24.7
Sulphur	1.8
Pyrites	16.8
Limestone	13.1
Chalk	9.1
Lime	5.0
Manganese oxide	6.2
Nitrate of soda	0.7
Products	
Soda ash	10.2
Bleaching powder	9.2
Sulphuric acid	2.8
Soap	1.8
Labour force 910	

Source: Pollution of Rivers in Scotland, 4th Report of the Commission on Pollution of Rivers, *B.P.P.*, 1872.

and some of them are as infernal in appearance as we can well imagine any earthly place to be. The buildings occupy an immense square from which shoot up numerous chimneys' (Tennant, 1922, pp.19–20). In competitive terms, however, all was not as well as appearances might suggest.

There were still, it is true, important local outlets for alkali—the 1863/4 post office directory for Glasgow lists 11 bottle-works, 21 bleachers, and 21 soap manufacturers—but demand was small in comparison to that available to Lancashire works. In transatlantic trade Glasgow had better shipping connections than the Tyne but was less well placed than Liverpool, where, indeed, Tennants opened an office in about 1830. St. Rollox was early so prominent in soda crystal manufacture that long afterwards its product was commonly known in London as Scotch soda, but for access to that market Tyneside was better placed. (Great Exhibition, 1852.) In terms of raw material supply the location was unfavourable. Coal was locally available, from Lanark pits either by rail or by canal, but was almost certainly usually more expensive than the small coals of Tyneside. For limestone the Clyde was less favoured than either Lancashire or the Tyne; in the 1860s most of the stone came from Ireland via the Forth and Clyde Canal. (*Eng.*, 28 Feb. 1868, pp.176–7.) There was no Scottish source of rock-salt, so that St. Rollox was dependent upon Cheshire. Interestingly, Tennants bought their way into the salt-mines discovered at Carrickfergus, but there is no evidence to suggest that these ever become a major source of their supplies. (*J.S.C.I.*, 31 Aug. 1889, p.655.)

There were also local problems. As the St. Rollox area became built up so waste disposal grew more difficult. The Pinkstone Burn and the Kelvin became polluted by the yellow liquid heavily charged with sulphur that ran from the works' 'deposit land'. (*C.N.*, 2 Oct. 1868, p.160.) In seventy years' operations to the mid-1890s St. Rollox turned out an estimated 2,000,000 tons of tank waste (Stuart, 1911, p.2). There were problems of gas disposal, and in 1842 Tennants went so far as to build a chimney $455\frac{1}{2}$ ft high. (*M.J.*, 14 Dec. 1850, p.590.) Both problems were eventually solved, but the disadvantages of operating on the fringes of a great city remained. Proof of the difficulties faced by the Scottish industry and in particular by St. Rollox is found in the subsequent history of the Tennant group and also later in the policy of the United Alkali Company.

There was in the first instance a strong movement of Scottish entrepreneurs and managers to other districts of Britain. MacBryde and Shank were important in the establishment of two St. Helens works, Hay Gordon and Neil Mathieson in Widnes. In 1864 Tennants themselves built a Tyneside works. A variety of factors induced them to take this step. A Tyneside plant was better placed to serve both the London and the Continental demand for alkali. (Tennant, 1922, p.19.) Land on Tyneside was in short supply and very expensive, but the site was open in comparison with that at St. Rollox. An important factor was the relative cheapness of coal on Tyneside, and another seems to have been the lower cost of labour. By 1877 the St. Rollox average annual wage was £59. 5*s*., that of Tennants' Hebburn works only £38. 3*s*. 5*d*. (Noxious Vapours, 1878, p.275.) A further indication of the advantages of Tyneside was the way in which the range of products was extended at Hebburn. When production began in 1864, Hebburn was designed to work up sulphate of soda sent from Glasgow, but in 1874 it began to make its own vitriol, to decompose salt, and to make bleaching powder. In 1877 the total production of saleable chemicals at St. Rollox was just over twice as great as at Hebburn. Forty-five years later, and within the United Alkali Company organization, St. Rollox's combined chemical capacity was 118,000 tons to Hebburn's 98,000 tons. St. Rollox survived until 1965. It was, however, but a shadow of its former self: in 1875 it had employed 2,000; when taken into I.C.I. it gave work to only 100. (Campbell, 1971, p.64.) In its last decades St. Rollox specialized in sulphuric acid, with a capacity in the late 1950s of some 40,000 tons, some 7,000 tons less than in 1911. (Third Statistical Account, Glasgow, 1958, p.298; U.A.C., 1911.)

CHAPTER 7

Tyneside Alkali: the foundations

A substantial review of the chemical trade in 1875 by a leading trade journal concluded with the words 'Tyneside will long continue to sway the sceptre of the alkali trade of this country' (*M.J.*, 20 Feb. 1875, p.191). In fact Tyneside was already past its relative peak and was soon to decline absolutely (Table 20).

Table 20. *The Tyneside district in the British alkali trade, 1852–94* (percentage of salt decomposed by British Leblanc works)

1852	42	1885	24
1866	45	1890	21
1874	42	1894	21

Sources: various.

The district's early prominence and later decline (more rapid than the national average) were associated with a host of factors—changes in supply of materials, in technology, in markets both at home and overseas, a declining quality of entrepreneurship, the competititon from other local industries for investment funds, and the quality of local labour. Fundamental to the whole evolution was the status of the prime trade of the district, the production and marketing of coal. From this the alkali industry and its local market industries alike sprang and changes in fuel supply affected the industry throughout its life. Cheap coal was Tyneside's most outstanding asset for alkali manufacture and its disappearance was a substantial blow.

Coal and salt

In 1822 a perceptive observer, in an extensive survey of the Newcastle coal trade, commented on the availability of fuel for other trades and on its role as a source of capital for them, and incidentally, referred to the multiplier effect.

> But the coal trade, though the staple, is far from being the only business carried on in Newcastle; so extensive and profitable a concern gradually produces an overflow of capital, which the easy access here to the sea naturally directs into the channel of general commerce, and the abundance of coal to the cultivation of arts and manufactures, a large share being, as usual, employed in providing for the wants of the increasing population of the town. Newcastle enjoys, accordingly, not only a considerable and flourishing foreign trade, but is moreover rich with various and extensive manufactories. (Constable, 1822, p.469.)

The alkali industry was early on the scene both as a coal-intensive activity and as a source of supply to the other industries of what was at the end of the eighteenth century one of the world's half-dozen biggest industrial agglomerations. At its peak, half a century later, it was to be carried on in almost thirty works on the river and, through its employment of as many as 10,000 men, to provide the direct support for perhaps 50,000 inhabitants of an urban/industrial population approaching one-third of a million. By the early 1890s, in a conurbation still rapidly expanding, employment in the alkali trade had shrunk to about half this level, and some thirty years later the industry was dead. Yet, even in the 1960s, some old Tynesiders could still recall working 'at the chemicals'.

The early alkali industry was unspectacular in the manufacturing economy of Tyneside. The Frenchman Faujas Saint Fond, breaking his journey in Newcastle on his way to north-west Scotland in the mid-1780s was fascinated by the industrial landscape:

> This beautiful river, the Tyne, is rendered highly interesting by the number and variety of manufactures carried on upon its banks. On one hand are brickfields, potteries, glass houses and chemical works for making ceruse, minium, vitriol etc.; on the other manufacturies in iron, tin and every kind of metal; machines for making brass wire, plate metal etc.'* (Saint Fond, 1907.)

The vitriol works were at this stage essentially copperas works—the first lead chambers for sulphuric acid manufacture were not built on the river for another twenty years. Copperas, or iron pyrites, were brought from the pits and allowed to weather over a period of three to five years into 'green vitriol' (ferrous sulphate) which was then boiled to produce sulphuric acid. There were copperas works at Walker, at Willington Quay, on the Ouseburn, at Elswick, Scotswood, Dunston, and Felling. Local production of sea-salt was declining rapidly in the decades around 1800 and the salt-pans had no link with the Tyneside alkali trade. Though they were very much less spectacular, brine springs in the coal measures were more significant. The discovery in 1796 of a brine spring at Walker colliery which had been sunk eight years earlier was of great importance and led to the establishment of the first Leblanc works in Britain. Even more important perhaps than coal and salt in the localization of the alkali trade on the Tyne was the demand from consuming trades established there directly because of the availability of coal. Haber (1958, p.17) has noted of Tyneside, 'the small local soap boilers and glass works could never sustain a chemical industry'; they did, however, perform the vital function of initiating its development.

Glass and soap

In 1616 Robert Mansell's glassworks had been set up in the Ouseburn near the present Glasshouse Bridge, the industry spread from there along the Tyne and also to the Wear and to locations on the Northumberland coast. In the

*Ceruse was a white paint made from carbonates and hydrates of lead; minium was red oxide of lead; vitriol was sulphuric acid.

decades around 1800 there was great expansion. (Middlebrook, 1950, p.141.) by 1828 there were twenty-eight glass houses on the Tyne (Pilbin, 1935, p.247; Mackenzie & Ross, 1834, p.ci). The district had some local materials notably its small coal available very cheaply, some fireclays (though Stourbridge clay was also used in the eighteenth century), and local sand—including shore sand at Hartley for Seaton Sluice bottle glass and Jarrow Slake deposits which, mixed with sea-water, produced the almost black bottles used in the local beer trade. Another mixture was Tyne sand, lime, and the refuse of the soap and alkali works. Sand from Alum Bay and from King's Lynn and also Belgian glass sands from the Moll district of Kempenland were brought in as ballast on returning coal boats. By the 1820s north-east England was the nation's prime glass-making district. (Lewis, 1831; Knight, 1860, p.398; Pilbin, 1935, *passim*.) The first soap-works were built in Westgate, Newcastle, in 1770 and by 1800 the trade was of considerable importance. By 1840 there were three large soap-works in Newcastle. (Fullarton, 1842.)

Entrepreneurs from both the glass and the soap trades played a significant part in the inception of the mineral alkali trade on Tyneside, the Cookson family being prominent among the former, Thomas Doubleday and Anthony Clapham among the latter. In the eighteenth and early nineteenth centuries alkali for these and other trades was derived in part from the local kelp, but this was killed off by the competition from Scottish and Irish kelp and by barilla imports. Doubleday and Easterby derived the alkali for their soap-works partly from the soapery waste salts, Leith as well as local works being an important source of supply. These 'soapers lees' were made up largely of common salt with some sulphate of soda included. In 1808 the firm began to make sulphate of soda by decomposing lees. By 1812, in order to supply the acid for this process, they had obtained plans for sulphuric acid chambers from Tennants in Glasgow and built a plant at Bill Quay, the first on the river—indeed, at that time there were no other lead chambers within 100 miles. (Mackenzie & Ross, 1834, p.125; Stevenson, Clapham, & Richardson, 1863, p.124.) Nine years later plumbers from St. Rollox installed a second sulphuric acid plant at Walker, which had already achieved the distinction of having the first alkali works on the Tyne.

Walker chemical works

In the late eighteenth century both Thomas Doubleday and William Losh tried to make soda ash from common salt, the former spending as much as £1,000 on experiments before giving up. The Losh family persevered and eventually succeeded; in the process their fortunes became involved with the Walker brine spring. On the Losh estates at Woodside a few miles south of Carlisle experiments in alkali manufacture were begun about 1790 with the co-operation of the Earl of Dundonald. By 1793 activities had moved to Bell Close on the Tyne, and it is suggested that it was from the nearby Lemington ironworks that the heat for evaporating soapers lees was derived. (Campbell,

1962, p.12.) Probably in 1796 (Pilbin suggests 1796, Campbell 1798) Losh and Dundonald moved to Walker. There they were able to avoid the heavy excise duty on salt by a commitment to mix soot and ground coal with the colliery's brine so that it could not be used as a source of domestic salt. They were thus able to produce salt for twenty to twenty-four shillings a ton at a time when, with the duty, the normal price was £20 a ton. In the early years Walker produced about 100 tons of soda annually from brine, in the years immediately following the Napoleonic Wars some 400 tons. (Jeans, 1877, p.84; Salt Duties, 1818, pp.351, 396.)

William Losh gradually acquired a wider knowledge of alkali processes. In 1791, less than two years after Leblanc's process had been announced, he studied in Paris under Lavoisier. He made a brief visit to France in 1802 during the peace of Amiens and in 1816 undertook another technical mission there. When the Leblanc process was first worked at Walker is disputed, but it was probably in 1816 immediately after his return. Not until 1821, however, were acid chambers operated there and production on a bigger scale begun. (Stevenson, Clapham, & Richardson, 1863, p.124; Armstrong, 1864; Clapham, 1868, p.38; and Lunge, 1880, p.306.) The repeal of the Salt Duties in 1825 was followed in the next few years by a major extension of alkali production to meet the expanding demand. Walker brine could not support this extension of production, and in 1825 the first Cheshire rock-salt was delivered on the Tyne for twelve to thirteen shillings a ton, or little more than one-third the cost at that time of salt-making from brine. Use of Walker brine stopped in the following year. Losh in 1799 made 10 tons of alkali, 121 tons in 1820, but by 1825 406 tons, and in 1830 968 tons. (Haber, 1958, p.17.) Walker works and the soaperies and glassworks were already becoming sources of capital, and entrepreneurial and managerial skills for the growth of a wider spreading Tyneside trade.

The spread of alkali manufacture on Tyneside

The century-old glass-making firm of Cookson in 1822/3 set up soda-works in the centre of South Shields using plans passed on to them by Doubleday (Stevenson, Clapham, & Richardson, 1863, p.124). A few years later Cooksons moved out of the town to build new works on the east side of Jarrow Slake, and in 1835 put up another works on an old brickfield at Westoe one-third of a mile away from the shore. This works does not seem to have survived long (Bell, 1849). Jarrow works suffered from much litigation in connection with pollution and was closed in 1843. It was reopened in 1844 by the Jarrow Chemical Company which was much more successful in coping with pollution.

A bottle-maker, Coultherd, along with John Allen, a chemist and druggist, had already built the Tyne's third Leblanc works at Felling Shore in 1827/8, and another went up near by, probably in 1830, in association with an earlier alum and Epsom salt works. In the 1820s Charles Attwood operated glassworks and soaperies in Salt Meadows, and built alkali works just to the east of this district. Clapham, one of the first users of Losh's alkali, had already moved his soap operations from the Ouseburn to a district in which complaints about nuisance would be less. In 1827 he acquired a site at Friars Goose, and there two years later began to make soap; by 1831 he was also turning out soda. To the east, Hugh Lee Pattinson, who had worked for Clapham in the Ouseburn, built Felling works in 1834 along a dene, but served by a colliery wagonway. At least two other works were built in the 1830s, the East Jarrow works with quays on the Don stream running into Jarrow Slake in 1836 and, three years later, just to the east, the Don Alkali Works. Thereafter the chronology of Tyne works is sometimes difficult to establish, but by the mid-fifties there was a string of at least seven works from East Gateshead to Bill Quay. A group of four or five located around Jarrow Slake spread into South Shields. Between Bill Quay and Jarrow was a stretch of three miles of river with no works; here the river banks were steep and it was furthest away from the two nuclei of soap/glass-works at Gateshead/ Newcastle and South Shields. In the early 1860s two works were built on this stretch: those of Sheldon and Nixon and of Tennant, the latter being the more important.

Apart from Walker, the industry was slow to become established on the Northumberland shore of the Tyne. However in 1847 John Allen's interest in the Felling Shore works ended with the expiration of the lease and the dissolution of the partnership, and much of the equipment was moved to new works at Wallsend which soon had the highest rate assessment in that town. (Pilbin, 1935, p.306.) By 1857 there were at least three other north shore alkali works, in North Shields, St. Anthony's, and St. Peter's (Fordyce, 1857). In 1862 John Glover, one of Pattinson's managers, built works at Carville; during the next nine years two more works were built, one, Rose Hill near Wallsend and the other, operated by Lomas of Jarrow, at Willington. In short, the north bank was largely developed as an overflow for the older works on the Durham shore; this was the case with at least four of the seven new works built there. Altogether there were six Leblanc works on the Tyne in 1830, ten by 1840, and some twenty-six by 1866. Two of these last were above the bridges at Dunston and at Blaydon (Stuart, 1932); all the rest were in the $7\frac{1}{4}$ mile stretch of the river between East Gateshead and South Shields (Fig.9).

Early Leblanc works beyond Tyneside

Leblanc works were built at various times in at least four north-eastern areas away from the Tyne—near Blyth, at Seaham, and on the Wear and the Tees. Blyth had some local outlets in earthenware- and glass-works. In the 1820s the Blyth works was the second in the North-East, using salt from sea-water and copperas, though later acid was brought in from the Tyne and from Glasgow. (Clapham, 1868, p.42.) There were at that time other works at Cambois Point. By 1830, however, the local salt, copperas, and earthenware production had ceased, and in the mid-1830s the two alkali works were sold by

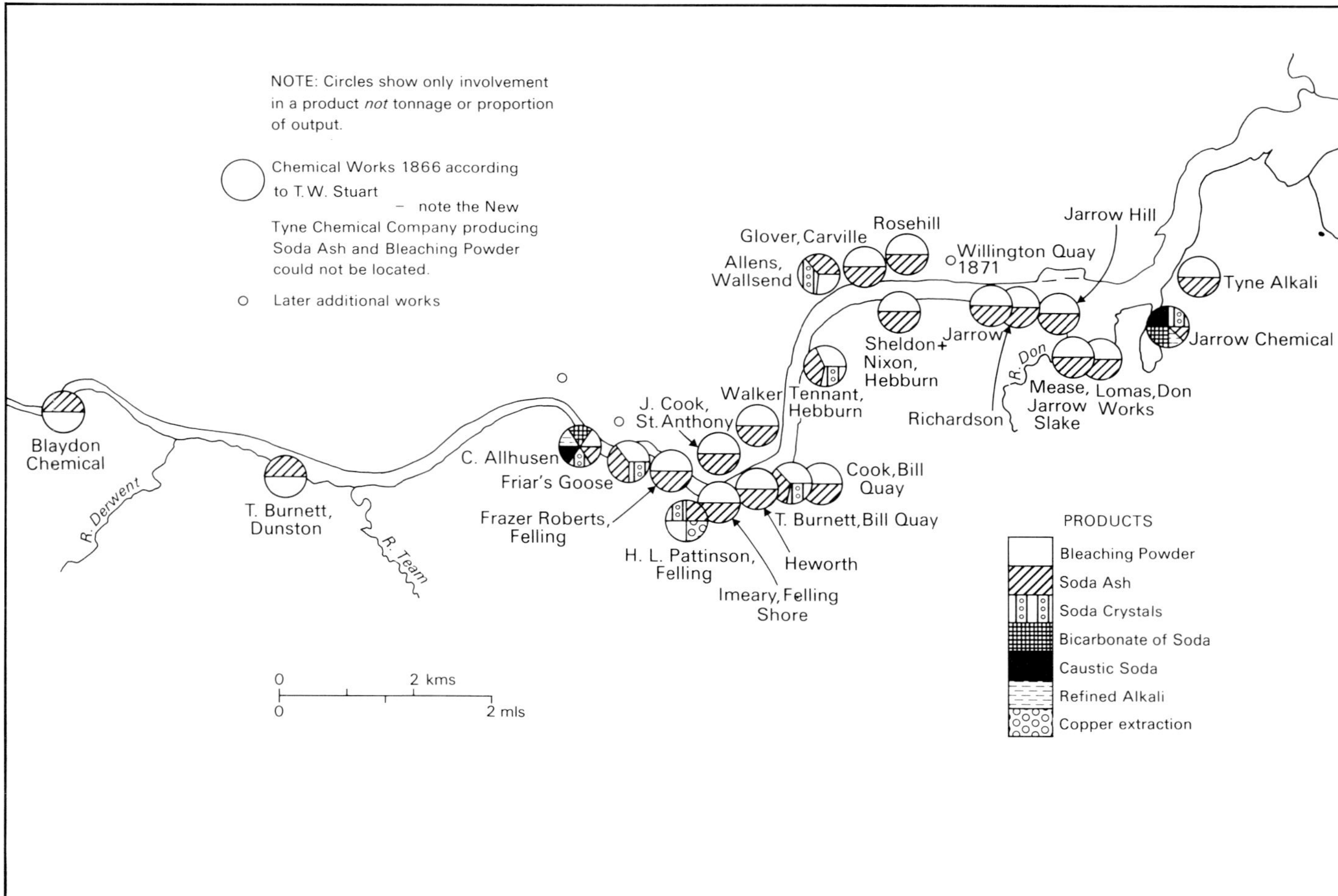

Fig. 9. Tyneside Alkali Works 1866 *Source:* Stuart, 1932

auction and seem to have been of no subsequent importance. (Lewis, 1831, 1848; *Newcastle Courant*, 20 Aug. 1836.)

Sunderland became a great centre of glass—particularly bottle—manufacture, with works strung out along the Wear from below the Wearmouth bridge to Pallion. By the mid-1850s sixteen bottle works made as many as sixty to seventy thousand bottles daily. (Fordyce, 1857, vol.2; O.S., 1857.) There were also paper-works along the river. In spite of his, however, Sunderland never developed an alkali industry. The reasons are difficult to isolate. That they may be associated in part with the lack of flat, broad sites owing to the deep incisions of the River Wear is suggested by the fact that the only major chemical focus to emerge in the neighbourhood was on the river well above the incised portion, at Washington.

In 1842 Hugh Lee Pattinson of Felling Chemical Works established works at Washington designed to make muriatic acid (hydrochloric acid) as its main product. Soda was a by-product. By the end of the forties it was already involved with magnesia production, and in the 1850s, when it was much enlarged, Isaac Lowthian Bell built a big plant for oxychloride of lead. (Lewis, 1831, 1848; *C.N.*, 23 Dec. 1904, p.316.) The works at Washington survived the elimination of the alkali trade.

Seaham Harbour was established in 1828 as a coal-shipping port. At the mid-century, in a mixture of natural industrial evolution and promoted diversification, new industries developed. From the 1840s there were bottle-works, and by the early 1860s John Candlish, one of the major Wearside firms, had seven bottle houses there. In 1865 the Seaham Harbour Chemical Works was built. It had an uneven career and closed in September 1890. (Whellan, 1856, p.643; *C.T.J.*, 24 Mar. 1888, p.188, 1 Oct. 1898, p.216; U.A.C. Tyne, various.)

The course of trade expansion and competition to the 1870s

Prices for chemical products fell sharply in the middle decades of the nineteenth century, but output grew almost equally spectacularly. The reasons are to be found in improved supply of materials, better production techniques, and larger scale of operation. Perhaps the most outstanding case was the expansion of the works which had been established by Attwood at Gateshead. In 1840 the works was sold to the grain merchant Christian Allhusen. Under Attwood operations had not been successful, but within a few years it became clear that Allhusen, though not competent technically, was an excellent business man. There was no intrinsic reason why Allhusen succeeded while others operated with much less reward. Assiduous attention to detail and a readiness to adopt proved innovations were the secrets. By 1861 Allhusens' raw material consumption varied from about one-sixth to one-fifth of that in all the chemical works on the river, with up to 600 ships discharging annually at its wharves. By the mid-seventies the works could deal with as much as 850 tons of salt weekly and by this time was producing just under one-quarter of all the chemical tonnage on the Tyne and gaining the consequent scale economies. (Fordyce, 1857, II, p.784; Jeans, 1877, pp.88, 130; *C.G.*, 17 Aug. 1861, p.105.)

Table 21. *Prices for soda ash and soda crystals, 1828–1872* (£)

	Soda ash	*Soda crystals*
1828	25[a]	20
1838	15	11
1852	10	5
1872	11	5

[a] In 1816 Losh sold soda ash at £60 a ton.

Sources: British Association, 1838, p.7; Richardson & Watts, 1863; *Great Exhibition*, 1852, p.23; *J.S.C.I.*, 29 Sept. 1882, pp.346–7.

Table 22. *Production by alkali works on the Tyne, 1838–1877* (thousand tons)

	Soda ash	*Soda crystals*	*Bleaching powder*	*Bicarbonate of soda*	*Caustic soda*
1838	10	20	Below 5	N.A.	N.A.
1852	23	43	5	4	N.A.
1862	43	51	11	7	0.6
1866	74	86	27	11	3.7
1877	90	117	30	7	1.5

Sources: British Association (1838); Richardson & Watts (1863); Lunge (1880).

Raw materials for Tyneside alkali manufacture

The Tyne was fortunate in possessing the most bulky inputs of the Leblanc process in abundance and cheapness; its trading connections put it in a uniquely favourable situation to assemble the others. In 1873 the North-East turned out less than one-quarter of a million tons of saleable products; its material inputs for this were 1.4 million tons. A deteriorating supply of materials was a major factor behind the district's worsening relative position in the industry from the late 1860s and goes far to account for its absolute decline which began in the second half of the 1870s.

For the production of sulphuric acid the Tyne was more or less as competitive as either Clydeside or Merseyside. Only a small quantity of nitre was used, and this was reduced by the adoption of the Glover tower. In 1862 the consumption on the Tyne was 2.5 thousand tons, or less than 1 per cent of all the raw materials for the industry imported into the river. In 1820 Tyneside made some 200–300 tons of sulphuric acid, by 1830 3,000 tons, and by 1867 140,000 tons. (Jeans, 1877, p.87.) Until 1840 all the acid made on the Tyne was derived from Sicilian sulphur—so that it was not uncommon at this stage of development to see as many as three or four ships from Girgenti discharging at one time on the river. (*C.G.*, 26 Oct. 1861, p.266.) However, the monopoly of 1838 caused delivered prices on Tyneside to go up from £8 to £14 a ton of sulphur. In June 1840, reacting to this challenge, John Allen imported the first consignment of pyrites. Others also installed pyrites burners, first using Cornish and then Wicklow pyrites, and continued in this even when Sicilian prices fell. By the mid-1860s Tyne works annually used about 70,000 tons of pyrites, and only a few hundred tons of sulphur, for special purposes. British pyrites supplies proved insufficient to make the home manufacturers independent of the wild fluctuations in the price of native sulphur (Stevenson, Clapham, & Richardson, 1863, p.125). Subsequent developments indicate the quick response of the Victorian capitalist to changes in supply and price. In 1860 Ireland was the leading source, supplying 4–5,000 tons. However, ballast rates on coal boats returning from Antwerp meant that sea freight on Belgian pyrites was as little as 6*d*. or 1*s*., or even sometimes as little as 3*d*. a ton of pyrites, compared with up to 7–10*s*. a ton from Wicklow. Imports from Belgium began in 1858, and from Westphalia and Norway three years later. By 1863 the Tyne imported only 4,000 tons from Ireland. (Kingzett, 1877, p.14; Jeans, 1877, p.86.) In the early years cupreous ores were avoided. When means of extracting copper had been devised, the residual ore was first sent to Swansea, but in 1865 copperworks were built at Hebburn and later also at Willington and Jarrow. Kingzett (1877, p.59) noted that the copper trade had reduced the cost of sulphur to the soda-maker by 40 per cent in twelve years. In salt supply the Tyne was, throughout its heyday and beyond, always at a grave disadvantage compared with Lancastria. Salt consumption in the Tyneside alkali trade rose from less than 50,000 tons in the late 1840s to 90,000 tons by 1862 and over 150,000 tons in 1866. For a period in the mid-fifties a small tonnage was derived from Carrickfergus, but, this apart, from the repeal of the Salt Duties in the 1820s onwards Cheshire was overwhelmingly the chief source. In the early years it seems that the salt was sent out via Liverpool and by coastal vessel around Land's End and the English Channel. By 1850 some Liverpool shipments went to Maryport whence the salt was carried by the Newcastle and Carlisle Railway to Tyneside, though this business seems to have been less important to Maryport than its import trade in cattle, timber, flax, or iron. (Knight, 1851; National, 1848.) Later, most was sent either by rail, or, more important, via canal from the Weaver to Goole, Grimsby, or Hull where it was picked up by colliers usually returning from foreign ports. The cost of freighting across country was substantial, but the charges from the Humber to the Tyne were nominal. In 1856 salt at the Weaver works was from 5*s*. 6*d*. to 8*s*. a ton, but at Hull the price was 17*s*. to 17*s* 6*d*. (Mineral Statistics of the United Kingdom, 1856, p.121.) Fourteen years later Cheshire rock-salt delivered in vessels in the Weaver was 3*s*. 9*d*. a ton. From there it was only a small charge to Runcorn or to Widnes. Salt sent by water from Liverpool to the Tyne went at a sea freight of about 6*s*.; the rail and canal freight to the Humber was 8*s*. 6*d*. (Forster &

Bell, 1870, pp.166–7.) Possibly the limitation of shipping capacity or of suitable return cargoes restricted the amount of salt that could take the sea route, though this does not seem a wholly satisfactory explanation. Soda was then selling at £11 a ton, and in the price of salt Lancashire works had the considerable advantage over the Tyne of 8*s.* a ton.

The alkali trade used immense quantities of limestone. In 1860/1, Peak limestone cost about 6*s.* 8*d.* a ton to Lancashire works; Newcastle works were then paying less than one-quarter as much for their limestone. (Richardson & Watts, 1863, p.318; Gossage, 1861, p.81.) As with salt, Tyneside benefited from the low ballast rates on returning colliers. These had for a long period been building up Tyneside's 'ballast hills' with sand and gravel dredged up from Woolwich Reach for their return journey from the Thames. They now found it more profitable to bring back chalk as ballast, mostly from Greenhithe or Northfleet. Although the cost advantage over Lancashire was substantial, the chalk was in many ways an inferior material, and much of it was heavily watered. In 1862 the Tyne used a substantial further supply of lime in the form of what was known as 'French cliff', a rather harder material brought by returning colliers from the outcrops along the Seine below Rouen or along the French coast. Unlike Thames chalk, this had to be ground before use but was preferred for bleaching powder production, though it was too dear to employ in the black ash furnaces. (Stevenson, Clapham, & Richardson, 1863, p.863; Lunge, 1880, pp.364, 365.)

In the mid-1840s the first sign of a threat to this system of limestone supply was signalled with the Wallsend launching of the ironbuilt collier *Q.E.D.*, the first vessel to be equipped with a double bottom for water ballast in place of the old solid ballast. In 1852 the threat became more obvious when Palmers at Jarrow launched the *John Bowes*, an iron screw-driven collier which cut the return journey to London to about one-sixth of the former level. Part of its equipment for reducing journey time was water ballasting, the water being let in as the vessel came down the Thames, and pumped out as it approached the Tyne staithes for reloading. Technical innovation thus helped north-eastern coal-mines to meet the competition in the London market of Midland pits shipping by rail, but its wide adoption helped to make the Tyneside alkali industry less competitive with Lancashire. (Armstrong, 1864, p.165; Lunge, 1880, pp.364–5.) In the late 1870s, when limestone cost 4–7*s.* a ton delivered at works in Scotland, the Midlands, or Lancashire, large chalk still often cost only 2*s.* 6*d.* at Tyne works (2*s.* f.o.b. Thames). So-called 'smalls' were only 1*s.* on the Thames and 1*s.* 6*d.* delivered. However, when winds were contrary or supplies for any reason scanty, large chalk delivered could range up to 5*s.* a ton. (Lomas, 1880, pp.157–8.)

As the decline of solid ballasting reduced the amount of chalk brought to the Tyne, the alkali firms turned to limestone. Much seems to have been railed from Buxton. By 1890 Fulwell and Whitburn between the Wear and Tyne were also important suppliers of Magnesian limestone, but Derbyshire lime costing three and a half times as much delivered was also used. (Campbell, 1962, p.19; British Association, 1889, p.146; U.A.C. Tyne, 14 Mar. 1891, 23 Mar. 1891, 20 July 1891.)

The material of which the largest tonnage was used in the Leblanc process was coal, the Tyneside industry employing about 0.32 million tons in 1862 and 0.71 million tons a decade later. It was in this sector that the district's greatest advantages lay. In 1860/1 fuel costs seem to have been about half those of Lancashire works (Table 23).

Table 23. *Cost of raw materials on the Tyne and in Lancashire, 1860/1*

	Tyne 1860		*Lancashire 1861*	
	per ton of material	*per ton of soda ash*	*per ton of material*	*per ton of soda ash*
	£. *s. d.*	£ *s. d.*	£ *s. d.*	£. *s. d.*
Sulphur in pyrites	7 17 9	2 10 5	5 0 0	1 15 0
Salt	16 4	19 2	8 0	10 0
Limestone	1 6	2 5	6 8	10 0
Fuel	2 11	10 0	6 0	1 1 0
		4 2 0		3 16 0

Note: The figures are not exactly comparable, those for Lancashire being less clearly defined.
Based on: Gossage, 1861, p.81; Richardson & Watts, 1863, p.318.

On Tyneside there were coal-pits on or very near the river and interspersed with the chemical works; there were good road or tramway and rail connections to landward pits. Moreover, 'small coal', which could not be exported or shipped coastwise with profit, was regarded almost as a waste product on the Tyne and was available at little more than the cost of haulage to manufacturers. In the early 1860s, however, supply was already shrinking and prices rising. Pits were flooded and abandoned at Wallsend, Willington, Howdon, Percy Main, St. Anthony's, Hebburn, and Jarrow. At the same time rail movement helped to prevent smalls from being a drag on the local fuel market; exports were increasing and general manufacturing demand going up. As a result the price of small coal increased fourfold in a few years. (Bell, 1862, pp.140–1; *The Times*, 14 Jan. 1865, p.7; *C.G.*, 10 Dec. 1864, p.464.)

Into the early 1870s coal consumption per ton of salt decomposed on Tyneside seems to have been about 3.09 tons as against 2.99 in Lancashire (Coal, 1871, p.200). Fuel economy now became desirable. During the seventies the introduction of mechanical operations helped. Before it disappeared, the district's advantage of cheap fuel was instrumental in causing one of the most important inter-regional shifts of chemical manufacture in Britain, that involving the establishment of Tennants' Hebburn operations. Less than fifteen years after its inception Hebburn was half as big as St. Rollox (Table 24).

Table 24. *St. Rollox and Hebburn works, 1877/8 Annual raw material consumption* (thousand tons)

	St. Rollox	*Hebburn*
Salt	29	15
Coal	127	55
Limestone and chalk	50	27
Pyrites	18	10
Brimstone	0.8	—
Nitrate of soda	0.4	0.1
Manganese	6.8	0.2
	232	107
Products	(thousand tons)	
Total weight excluding waste and burnt ore	37	18
Burnt ore, coal ashes, and alkali waste	76	40
Employees	1350	1000
Wages (£000)	£80	£38
Area occupied by works (acres)	50	30
Deposit land (acres)	80	21

Source: Noxious Vapours, 1878, p.275.

CHAPTER 8

Merseyside Alkali: national primacy and intra-regional locational change

In the second quarter of the nineteenth century Lancastria (Lancashire, Cheshire, and the Dee Estuary) was a major focus of alkali manufacture; in the third quarter it became pre-eminent. Yet, as on Tyneside, until a late stage chemicals did not bulk large in the regional economy. Demand for bleach for textiles; the Merseyside soap trade; the glass of St. Helens, Manchester, and Salford; the major paper-making of the Manchester region—all were important and growing outlets. (Ashmore, 1969.) Glass production here does not seem to have increased much in the first decade after Waterloo but did so rather later (Barker, Dickinson, & Hardie, 1972, p.144). National hard-soap output went up 30 per cent between 1815 and 1825. Cotton textile manufacture more or less doubled in the same decade. (Lewis, 1833; Hardie, 1961, p.6.) In the forty years to 1845 the population and the economic activity of Lancashire tripled.

In addition to its major home markets Lancastria was admirably placed to supply the international trade. Early Victorian Liverpool was, in the words of the secretary of its Stock Exchange, 'the greatest thoroughfare in the world'. With justifiable pride, and a not-uncharacteristic touch of Victorian hyperbole he developed his case, 'Her river is the great highway of nations: tide after tide, as it rises, brings into her spacious docks ships richly laden with this world's treasures,—the produce of every clime; and tide after tide, as it recedes, carries on its ample bosom ships fully laden with manufactured goods, prized and valued in every market in the world to which they have gained access.' (Scrivenor, 1849, pp.2, 3.) Liverpool was especially advantageously placed for the American trade. In the course of the century prices for alkali in Liverpool dominated the business, and at the end of the century it was a board of directors sitting in Liverpool which controlled the last years of the demise of the Leblanc industry. In short, from both a domestic and an international point of view, Merseyside was as favourably placed for alkali marketing as could be conceived (Table 25). To these advantages were added those of raw material provision.

Apart from sulphur (or pyrites) and nitre, all the chief material inputs for the Leblanc alkali process were abundantly available within or on the fringes of Lancastria. As late as 1873, Lancashire, Cheshire, and North Wales produced almost one-sixth of the coal of the United Kingdom (20.5 million tons), ranking second only to the 29,000,000 tons of the North-East, and well ahead of all other districts. By the 1850s the worked fields of Lancashire extended roughly to the Manchester to Liverpool railway. Less than 15 miles

Table 25. *The Lancashire alkali trade, 1852 and 1866* (thousand tons)

	1852	*1866*
Soda ash	26	87
Soda crystals	3.5	25
Bicarbonate of soda	1.2	6.4
Bleaching powder	1.2	11.2
Caustic soda	N.A.	20.0
Number of men employed	1519	

Sources: Richardson & Watts, 1863; *B.P.P.*, 1867/8, XVIII.

to the south-east was the Cheshire salt-field. To link the two there had already been developed natural, improved, or largely artificial means of water carriage. The improvement of the Weaver had begun in 1720, in the 1750s the Sankey Canal was dug from the coal district around St. Helens to an outfall into the Mersey at Fiddlers Ferry south-west of Warrington. Large-scale movement southwards of coal to the Mersey, to Liverpool, and to the salt-pans of Cheshire was counterbalanced by a growing movement of salt down the Weaver. Finally, fringing the great synclinal structure of the Cheshire Basin within which were situated both coalfield and salt-field were the carboniferous limestones of Flintshire and of Derbyshire. Given the variety of sources of supply, the changing tonnages, and freight rate developments, there was no clearly dominant location for the industry, but the material balances did suggest either a coalfield or an intermediate location oriented on a transport focus (Table 26; Fig.10).

Table 26. *Material inputs per ton of soda ash South Lancashire, 1861, 1872, 1888* (cwt)

	1861	*1872*	*1888*
Pyrites	24	$17\frac{1}{2}$	$14\frac{1}{5}$
Saltpetre	1	$\frac{1}{2}$	$\frac{2}{5}$
Salt	25	29	24/25
Limestone	30	35	24
Coal or slack	70	88	60
Total	150	170	122.6/123.6

Note the anomalous 1872 figures.

Sources: 1861: Haber, 1958; 1872 and 1888: Mond, 1872 & 1888.

It seems reasonable to assume that with experience the true balance of locational advantage was more fully recognized, and that, as competition became keener, and profit margins, though long generous, narrowed, choice of location, like care and economy in processing, became more critical for commercial success. The occasional exceptions to the general trend may be regarded as mere quirks of circumstance, fascinating for what they reveal of the decision-taking process but only variations on the major themes of locational change. Nevertheless, they affected local economies, societies, and landscapes, and to this extent even the anomalies contributed to the formation of the economic, social, and urban geography of late twentieth-century Lancashire.

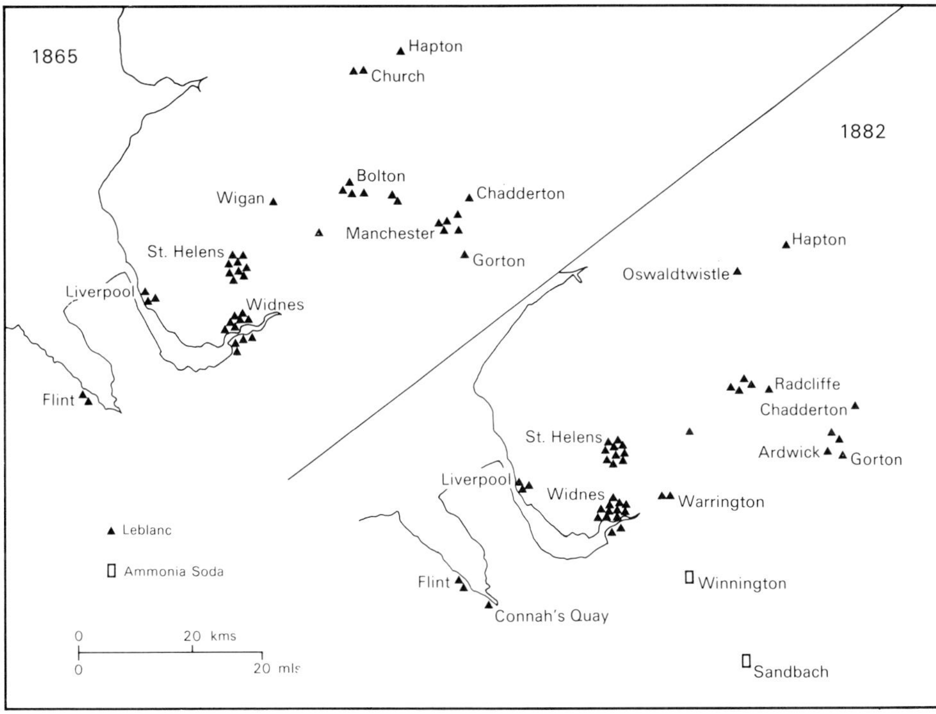

10. The Lancastrian chemical districts 1865 and 1882 *Source:* Alkali Inspectors' Returns

Liverpool pioneering in alkali

In the early development of chemicals in Lancashire Manchester played some part. About 1830 it had 'several laboratories for the making of oil of vitriol and other chymical productions used in the different processes of the trade, for bleaching dyeing and the like' (Lewis, 1833). Yet at the end of the forties chemical works employed only 66 of the 6,036 horsepower reported for steam-engines in Manchester and Salford. By 1851 sixteen alkali works were reported for Lancastria, two each in Blackburn, Wigan, and Bolton, two in the Runcorn area, and eight in St. Helens (*M.J.*, 15 Mar. 1851, p.129).

From 1814 in Dublin James Muspratt had made muriatic (i.e. hydrochloric) acid, acetic acid, turpentine, and later Prussiate (that is Prussian blue, a deep blue pigment made from a salt of iron). He contemplated Leblanc alkali, but could not command the capital, and recognized that Dublin was an unsuitable location, though the high prices then ruling might have covered the burden of even a grossly sub-optimal location. In 1822 Muspratt leased old glassworks on Vauxhall Road Liverpool with access to the Leeds and Liverpool canal. Here again he at first confined his operations to Prussiate, but in 1823, at the time of the reduction of the Salt Tax from 15*s*. to 2*s*. a bushel, he used his profits to build lead chambers for the production of vitriol and then went on to make Leblanc alkali. In this way, though anticipated both by Losh on the Tyne and by Tennant at St. Rollox, Muspratt became the first British producer to use Leblanc's process on a considerable scale. The coincidence of his entry with the reduction of the Salt Tax has not unreasonably been seen to represent a causal connection, though Hardie (1950, pp.16, 17) has suggested that it was merely a breakthrough against the established position of barilla and of kelp alkali and against the preferences of the soap boilers for them and that 'the coincidence of the dates of repeal [that is 1825] of the tax on Salt and the opening of Muspratt's Vauxhall works *is* a coincidence and nothing more'. It is true that Muspratt persuaded the soap-makers—of which there were seven or eight then in Liverpool—to take his soda ash rather than the potash which they had been accustomed to use by giving away the first tonnages he made, but the sharp reduction of salt costs must have helped to make such a promotional exercise possible. (*J.S.C.I.*, 29 May 1886, pp.314–15; U.A.C., 1923.) Repeal and chemical salesmanship probably both played significant roles. Within two years a second Leblanc works was operating in the same area under the control of the former acid-makers Lutwych & Co. (Baines, 1825, p.368; Hardie, 1950, p.12.) After 1830 Liverpool declined in relative importance as an alkali centre and thirty years later the industry did not secure listing among the city's main industries (Fullarton, 1859). As late as 1881 there were three alkali works in Liverpool.

Among Liverpool's assets the favourable marketing situation must rank high. For the export trade there were road and water connections directly to the docks about one mile away from the Vauxhall Road works (Hyde, 1971, p.40). There were local glassworks and soap-works and easy access to those of St. Helens (glass) and Warrington (soap). Liverpool was advantageously placed for deliveries to textile trades both east and west of the Pennines via the Leeds and Liverpool canal and the Liverpool and Manchester railway completed in 1816 and 1830 respectively. For the flats which carried the salt from Weston at the outlet of the Weaver navigation the journey to Liverpool docks was only a few miles longer than that up the Mersey to Warrington or to the mouth of the Sankey Canal at Fiddlers Ferry. Coal could be delivered direct to the Vauxhall Road from the Wigan area by canal or from pits near St. Helens either by rail or by rail and water. Finally, Liverpool was an important reservoir of both investment funds and entrepreneurship.

Yet Liverpool in the early nineteenth century (and indeed more than a century later) was primarily a commercial rather than a manufacturing city. Its preoccupation with trade, exchange, finance, and shipping made it unsuitable in some ways for the more regular routines of manufacturing. Two aspects of this can be seen in its attitude towards labour and its reaction to the nuisance occasioned by certain industrial activities. Baines (1825, p.195) suggested that the city's lack of immediate sources of coal would itself constitute a 'substantial handicap . . . were not the habits and pursuits of persons residing in commercial cities incompatible with that close confinement to which the manufacturing classes submit without reluctance'. Fifty years later Wilson (1872) found similar explanations for the fact that in Liverpool 'manufactures are, in a chief degree, either repelled by commerce or subsidiary to it'. These conditions soon forced Muspratt to look outside Liverpool.

Even before the Vauxhall Road alkali works had been established city opposition to noxious industry had led to the removal of copper smelting. Although neighbours of the alkali works put up with its nuisance when it was small, they began to complain as output increased. The correspondent of the *Liverpool Mercury* complained in October 1827 that the gas clouds were so great that 'the church of St. Martins in the Fields, now erecting cannot be seen from the houses at about 100 yards distance . . .' (Hardie, 1950, pp.226, 227). Muspratt was indicted by the corporation for nuisance and verdict was given against him. Lack of room for expansion was soon another problem.

Moreover, as Liverpool seemed less attractive, new considerations improved prospects elsewhere. Soap- and glass-making grew more rapidly to the east and, as prices fell, savings on the delivered price of coal became more important. By 1839/40 when coal slack was 5*s*. a ton in Liverpool, it was only half that in Newton in Makerfield. As coal consumption was anything up to 4 tons per ton of soda ash, this difference alone could mean as much as 10*s*. on soda ash then priced at £15 a ton. (Muspratt, 1886, p.410.) In 1838 Vauxhall Road works decomposed 12 tons of salt a day and employed 120 persons. After 1851 it ceased to conduct the full Leblanc process, though even in the late 1860s it employed 130–50 men and made 5,000 tons of black ash annually; it did not close until 1896. (Hardie, 1950, p.210; Barker & Harris, 1954, p.283 footnote.)

Newton and Flint

Within five years of establishing works in Vauxhall Road, Muspratt was considering operations elsewhere. The new growth points were along the Sankey Canal, tapping the Lancashire coalfield and giving direct if roundabout access to Weaver Basin salt. Moreover, by a provision of the Act for the canal in 1755, inserted with the intention of benefiting farming interests, carriage of limestone was permitted toll free. (Barker & Harris, 1954, p.228.) This area gained a new nodality in the early railway age; the Liverpool and Manchester railway crossed the Sankey Canal about half-way between the two metropolitan focuses of the region. In 1829 the Warrington and Newton Railway Company was incorporated and later built on northwards to Wigan. A little to the north of this canal/railway junction area there were coal-pits near West Leigh, Ashton, Haydock, and Black Brook. To the south were the important glass- and soap-works of Warrington, and westwards good access to St. Helens glassworks. It was in this strategically located middle ground at Earlestown near Newton that Muspratt built in 1830. But Newton and Earlestown became dominated by other industries—by the late forties

Muspratt employed 300 there, while two other large works, making locomotives and working iron, employed 700–800. The surrounding land was highly cultivated and, particularly to the south around Winwick, there was a good deal of woodland with which alkali fumes could play havoc. Muspratt was implacably pursued by local land-owners (Barker & Harris, 1954, pp.235–9); by 1849/50 the decision to move had been taken, and two years later Newton was closed and pulled down.

Muspratts now turned to Flint, where the town population was small (3,300 in 1851) and the firm's arrival could be seen as providing salvation for the local economy which was suffering from silting of the harbour, from the movement of county business to Mold, and from the closure of local firms. Moreover Flint already had acquaintance with the alkali trade; its chemical production had grown out of lead smelting. In 1852 Muspratts paid £7,000 for Roskells' old alkali works at Flint, which they rebuilt with the object of condensing muriatic acid gas as perfectly as possible and subsequently operated with success. (Imperial, 1872; Fullarton, 1859; Richards, 1973; Noxious Vapours, 1862, p.175.) By the early 1860s Flint produced about 11,000 tons of bleaching powder and soda ash annually.

In addition to its isolation, scope for waste disposal, its water supply, and absence of active pressure groups, Flint had other attractions. Coal-mines in the vicinity were able to supply the 45,000 tons of coal needed annually in the early 1860s, and limestone was quarried in the hills immediately behind the town, though later Buxton limestone was also delivered by rail. (Richards, 1973.) A new wharf was built to handle vessels of up to 300 tons, and these brought sulphur and pyrites to the works and handled its soda exports. By 1861/2 the tonnage of salt decomposed at Flint seems to have been twice that at the Muspratt family works in Widnes (Hardie, 1950, p.38). Other alkali works were built along the Dee estuary at Pentre by Smith and Maudsley (*c.* 1860), at Queensferry (1863), and at Connah's Quay (1881). Remoteness from industrial centres long preserved for Flint the advantage of lower labour costs; moreover unionization was slow—it was not until the early 1890s that a meeting was held to form a branch of the Chemical and Copper Workers Union (*C.T.J.*, 25 July 1891, p.49, 1 Aug. 1891, p.58). In the last few years of the nineteenth century, the trade began to decline and serious unemployment resulted. By 1911 chemical employment in Flintshire was only one-third the level of 1891. Muspratts' works, the last in the whole Dee belt, closed in 1920, and two years later the site was sold to Courtaulds. (Richards, 1973; *Encyclopaedia Britannica,* 1911).

Runcorn and St. Helens

In 1816 Thomas Hazelhurst began to make soap and resin along the Bridgewater Canal in Runcorn. During the next five years another works (which eventually became the Runcorn Soap and Alkali Company) was built for the same trades. In the thirties both works began to make vitriol and alkali. By then there was another works in Weston, and yet another went up in the late fifties. By 1890 there were three important alkali works in Runcorn and one at Weston. (Nickson, 1887; Norman 1931.)

St. Helens was the regional prototype of the chemical town; Widnes was to be the fully-fledged development. The first St. Helens alkali works was put up in 1828/9; within nine years there were four; by 1862, following rapid expansion in the sixties, seven or eight; by 1876 the town and neighbourhood had fourteen alkali and three other chemical works. Eleven alkali works survived in 1890. (Noxious Vapours, 1862, p.21; Worralls, 1876; U.A.C. *c.* 1930; Fig.11.)

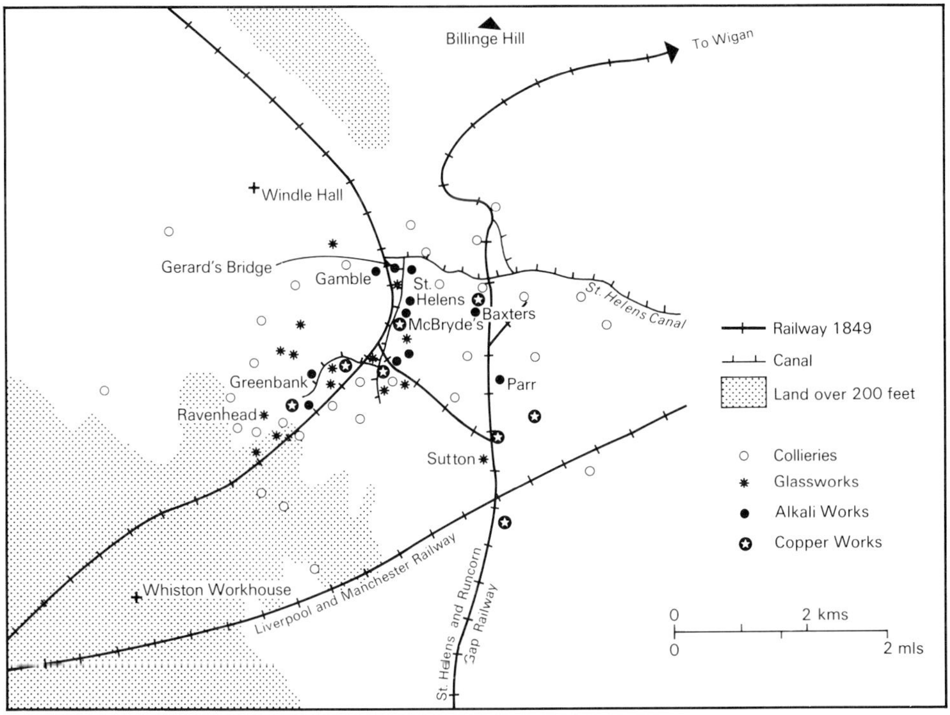

Fig. 11. St. Helen's alkali and other works Based largely on Barker & Harris, 1954

St. Helens lies on the exposed coalfield and in the second quarter of the nineteenth century had active pits in the immediate vicinity. In the early sixties these pits raised 2 million tons of coal a year, of which about one-third was used in the town. The Sankey Canal and its 1833 extension facilitated access to Weaver salt. The town, Barker and Harris (1954, p.228) suggest, was just about the ideal size: big enough to provide housing and services for workers, but not yet big enough to restrict the industry by means of litigation.

The population in 1830 was approximately 6,000, and sensibilities about landscape and amenity had already been softened by coal-mining, copper-smelting, and glass-making.

By 1829 St. Helens had only five glassworks—very few compared with the Tyne—but Ravenhead works at Sutton was the largest plate glass works in England. In 1840 or thereabout there were two other St. Helens plate glass works, and in Eccleston three works making crown and flat glass and bottles. By the late nineteenth century when there were nine works, Pilkingtons had already achieved a national pre-eminence. (Lewis, 1833 and 1848; Muspratt, 1886, p.413; British Association, 1896.)

The first St. Helens alkali works was begun in 1828 by Muspratt and J. C. Gamble at Gerards Bridge, the head of the Sankey Canal. (It is said that, before deciding on St. Helens, Gamble considered establishing an alkali works near Birkenhead Float, using brine pumped over 24 miles from Northwich (Allen, 1907, *passim*).) There were coal-pits on Cowley Hill and at Rushy Park no more than half a mile away; salt and toll-free limestone came directly up the canal. Many more St. Helens works promoters moved in from other districts. In 1836 Crosfields, with twenty years' experience in their Warrington soap business, acquired an interest along with Gamble in another Gerards Bridge works which had been set up in 1829 to make alum and subsequently converted to alkali. (Ashmore, 1969, p.131; Barker & Harris, 1954, pp.231. 232, 235.) Six years later Andreas Kurtz, a chemical entrepreneur who had already operated in London, Dublin, and Manchester, moved into St. Helens to take over and expand a small, failed concern (Allen, 1907, pp.117–39).

The growth of St. Helens was sometimes due to evolution from other lines of business, sometimes, as so typical of Victorian industry, to managers or other supervisory staff leaving established works in order to set up firms of their own. Baxters was a copper-smelting plant, but about 1872 it began to make caustic soda from bought saltcake using waste heat from copper furnaces for heating the liquors. It survived until 1927. James McBryde, who came to work for Gamble in 1847, broke away to build at Pocket Nook. Not all works operated the full Leblanc cycle of production—Thomas Walkers' works, established in about 1873, for instance, made only caustic soda using bought saltcake.

By the mid-1840s St. Helens was a town of approximately 12,000, and, as its growth secured it greater standing as a market town, opposition to pollution began to increase, opposition that was carried further by local land-owners led by St. John Gerard, lord of the manor of Windle, whose hall was only just over a mile north-west of Gerards Bridge. Within twelve years of the setting-up of the first St. Helens works, pollution was so serious that Gossage's recovery towers were being built in the district. (Barker & Harris, 1954, pp.236, 237.) A more decisive check to the growth of St. Helens was administered by changes in transport facilities and in freight rates.

In 1845 the St. Helens and Runcorn Gap Railway Company and the Sankey Brook Navigation Company amalgamated to form the St. Helens Canal and Railway Company. In the next few years this company improved and extended its rail system, with substantial overhaul and modernization of the line to Runcorn Gap, the opening of a branch from there to Garston in 1852, and new docks the following year. A railway from Runcorn Gap to Warrington was also opened in 1853. As a result, a new importance was conferred on the small community of Widnes. The changes in physical accommodation for transport were accompanied by even more decisive changes in freight-rate policy. Progressively over the years the freight rates on carriage by both canal and railway were raised by the St. Helens Canal and Railway Company, which now had a local monopoly. A Widnes rather than a St. Helens location looked especially attractive, for the rise in rates affected all commodities except coal. (Barker, 1960, p.130.) The first rate increase came in the summer of 1848; there was a sharp upward revision in October 1854, and altogether from 1845 to 1854 the rates on bulk cargoes rose from about 8*d*. to 1*s*. a ton by canal and doubled by rail. In 1864, in evidence on the St. Helens Canal and Railway Transfer Bill, David Gamble stressed how this penalized St. Helens in comparison with Widnes: 'What I complain of is that practically our distance from the raw materials has been increased by the increase of these rates whereas their distance has not been increased at all'. (Barker & Harris, 1954, p.343.) Less spectacularly, fuel economy was reducing the advantages of a coal-oriented location and consequently increasing the attractions of sites closer to other raw materials.

However, pollution, new lines of transport, and changed freight rates did not stop the growth of St. Helens alkali. Powerful momentum for further growth was derived both from a mixture of external economies and from limited perception of the changing circumstances on the part of entrepreneurs. In the years 1852–4 three new plants were built, at least two of them by former employees of other St. Helens firms. Barker and Harris (1954) show that the large increases in freight rates of October 1854 cut out all construction of new works for ten years, but established works expanded. Gerard's land agent reckoned that between 1852 and 1862 Gamble's works doubled its capacity, Crosfields increased by 50 per cent, and Radley chemical works increased fourfold. (Noxious Vapours, 1862, p.9.) In 1864, when the St. Helens Canal and Railway Company was taken over by the London and North Western, alkali-makers were granted rate reductions—down from 1*s*. a ton to 10*d*. a ton by canal and from 2*s*. to 1*s*. 8*d*. a ton by rail. This was followed by some revival, and even two new works. There was some expansion of existing works, and others, when put up for sale, found buyers. The last St. Helens chemical works survived until the late 1920s, but from 1860 growth at St. Helens was overshadowed by that at Widnes.

Widnes

In mid-1833 both the St. Helens and Runcorn Gap Railway and the newly extended Sankey Canal were opened to the low ground at Runcorn Gap. Coal was regularly shipped out from the dock from then on. Until the 1840s Widnes was part of Appleton township and of so little significance that its name was still spelt variously—Wildness, Widness, or Widnoss. (Arrowsmith, 1815; O.S., 1842; Board of Trade, 1845, p.383; Fig. 12.) By the late 1840s its transport facilities were being greatly improved and in terms of transport costs its situation was excellent for bulk industry. The attention lavished by the St. Helens Canal and Railway Company on improvements to the railway from St. Helens released land along the canal at Widnes which had formerly been earmarked for dock extensions. (Barker & Harris, 1954, p.343.) In August 1848, a few weeks after the first freight rate increases were announced by the St. Helens Canal and Railway Company, John Hutchinson of St. Helens applied for a lease of land on the canal bank near the outfall into the Mersey for the first Widnes alkali works. The fact that this was only seven miles from St. Helens permitted Hutchinson to attract trained labour, particularly furnace men, from the older centre.

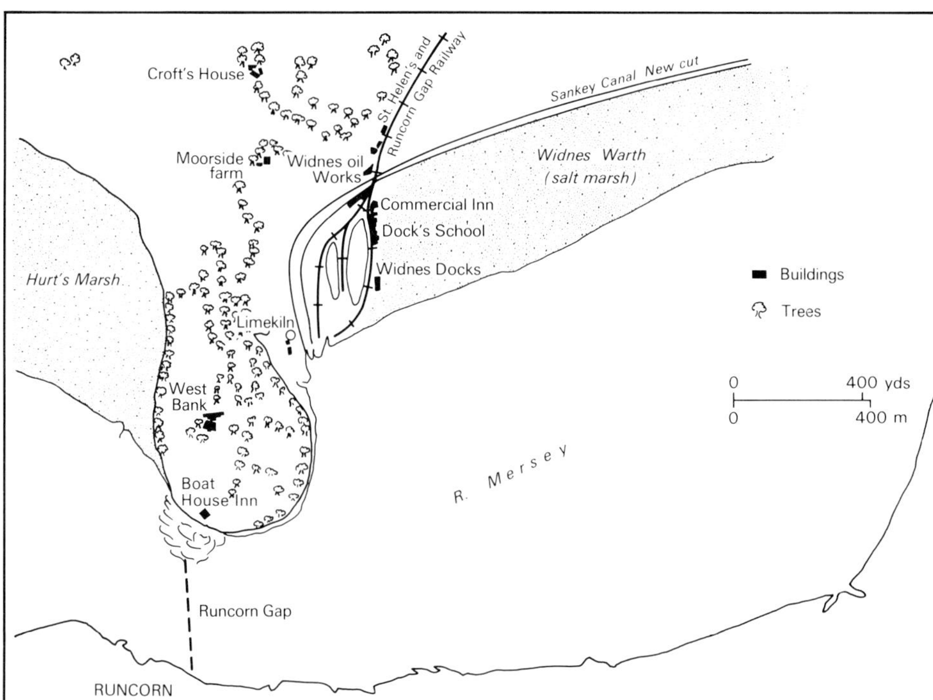

Fig. 12. Widnes district 1847 Based on Ordnance Survey, Six Inch Map, surveyed 1846/7

Hutchinson, a native of Liverpool, had moved to St. Helens to work in Kurtz's works. When Kurtz died in 1846 Hutchinson was in charge, but a clause in Kurtz's will required the mortgaging of the plant. In these circumstances Hutchinson decided to seek a more secure situation. Fourteen years after he started to build in Widnes, he spoke to a parliamentary committee of the operations of his works, by then extended by the construction of a second plant just across the canal. He now made 15,000 tons annually of soda ash, soda crystals, and bicarbonate of soda. For every ton of product he assembled over six tons of material (Table 27).

Table 27. *Hutchinson's Widnes works, 1861/2 annual production and supplies* (thousand tons)

		tons per ton of product
Saleable products	15	
Materials consumed		
Salt	14.5	0.97
Pyrites	10–15	say 0.83
Limestone	20	1.33
Coal	50	3.33
		6.46

Source: Noxious Vapours, 1862, p.169.

When the absence of canal toll charges on limestone and the fact that pyrites could be railed from Liverpool docks to St. Helens are taken into account, these figures suggest that a coalfield location remained more desirable for alkali manufacture than an intermediate one such as Widnes. But there are a number of qualifications. In the first instance, as limestone quality became more important, it seems probable that it was railed from North Wales and the Peak, thus giving Widnes a slight mileage advantage over St. Helens. (For Hutchinson limestone was possibly even cheaper, for from 1853 to 1864 he was in partnership with Oswald Earl who had interests in lime (Hardie, 1950, p.28).) In delivering alkali to export points Widnes again had a slightly shorter rail haul and direct access to Garston docks. Taking Hutchinson's 1862 tonnages, the nearest major source of supply, and the length of journey either from this place of origin or, alternatively, from a common point of dispersal to the various possible alkali-manufacturing locations, one can get a reasonable indication of the distance and ton mileages for each centre of production. The balance of advantage clearly lies with Widnes, emphatically so as compared with a hypothetical salt-field location at Northwich, to a smaller but not insignificant extent in comparison with St. Helens (Tables 28 and 29).

Table 28. *Transport distance for the supply of raw material and delivery of products from Lancastrian works, 1861/2* (miles)

	Northwich	*St. Helens*	*Widnes*
Limestone			
Clwydian (rail from Flint)	46½	45	38
Buxton (rail from Altrincham)	13	23	16
Coal			
From St. Helens by rail	27½	—	7
From Wigan by rail	46½	12	19
Salt from Northwich (by water)	—	32½	19
Pyrites from Liverpool (by water)	28	28½	15
Finished product to Liverpool (by rail)	32½	19	12

Based on: Noxious Vapours, 1862.

Table 29. *Lancastrian works, assembly and distribution for a production of 15,000 tons alkali annually* (ton miles)

	Northwich	*St. Helens*	*Widnes*
Limestone			
(a) Clwydian Range	930	900	760
(b) Buxton	260	460	320
(c) Assuming half *a* half *b*	595	680	540
Coal			
(a) All from St. Helens	1375	—	350
(b) All from Wigan	1850	300	650
Salt			
From Northwich	—	471	275
Pyrites			
From Liverpool	350	356	187
Finished products			
To Liverpool	487	285	180
Total			
(a) Most favourable combination	2472	1572	1312
(b) Least favourable combination	3817	2312	2052

Based on date in Tables 27 and 28.

Freight rates rather than ton miles, however, are significant. In general Widnes might be expected to lose in this respect, for it was an intermediate and not an originating point for any raw material and therefore would appear to gain no advantage from the tapering of rates with distance—if such tapering occurred on generally short hauls. On imported raw materials or on shipments Widnes had from the first the advantage of a competing medium, by water, which should have moderated rail rates. Later it gained from its direct access to three different railway systems, the London and North Western, the Cheshire Lines, and the Sheffield and Midland Joint Railway (a short local system formed in 1869 and operated jointly by the Midland and the Manchester, Sheffield and Lincolnshire). By contrast, not until the late 1890s was the London and North Western monopoly of St. Helens railway traffic broken by the arrival of the Great Central system. There is evidence, albeit scanty, that Widnes gained from this competition. The London and North Western opposed the coming of the Sheffield and Midland line and in response the latter was active in providing improved access to the Widnes works. (Widnes, 1872.) By the beginning of the 1880s, the nominal rate on Lancashire coal to Widnes was 2*s*. 6*d*. (though export coal came at 1*s*. 2*d*.). Even so the Muspratt works at Widnes paid only 6*d*. carriage for most of its coal. At that time hardly any of its outgoing traffic went by rail, but by water to Liverpool. (Muspratt, 1881, pp.142, 153, 762.) When the London and North Western asked for 1*s*. 4*d*. for one freight haul, Gaskell Deacon refused to pay more than 10*d*., stuck out for this as a fair rate, and did in fact pay only that (Muspratt, 1881, p.141). The Hendre Limestone Company north-west of Mold had sent limestone to Widnes, but rates were very high compared with those from the Peak District (ibid., p.140). It is perhaps significant in this connection that the Hendre quarries were in a locality in which the London and North Western held a monopoly, whereas various Peak quarries could be served by either the Midland or the London and North Western. Use of the Sankey Canal continued for many years—as late as the first decade of the twentieth century it was used for about 4,000 barge deliveries to chemical and soap works in Widnes. However it declined rapidly after that: in 1919 only seven flats passed through Newton locks on their way to St. Helens, and the canal became completely derelict that year. (Widnes Guide, *c*. 1910, p.18; Millward, 1955, p.111.)

Widnes had other advantages. As it was a mere village it was free from harassment on grounds of nuisance for some years at least. Even then, being essentially the creation of the alkali trade, and increasingly dependent upon that trade, pollution was commonly regarded as the unfortunate but necessary price paid for the subsistence of its population. Sir Edward Watkin, Chairman of the Manchester, Sheffield and Lincolnshire Railway, expressed this bluntly to the committee considering the Widnes Railway bill in 1872 (Widnes, 1872, pp.14–16): 'These traders have been turned out of different parts of England because they destroy the vegetation. The trade can only settle itself in certain districts, and it is for the interests of public health and everything else that it should be confined to certain districts.' Other external economies were derived from the large volume of rail traffic involved: in the early 1870s the inward rail movement to Widnes works—overwhelmingly to chemical plants—was 0.8 million tons and their outward traffic 0.37 million tons (Widnes, 1872, p.16). There were benefits associated with service and supply trades—of eleven non-alkali works in the town by 1872, at least eight were functionally closely connected with that business (Widnes, 1872, pp.14–15). Widnes, in short, was the most characteristic chemical town—'Leblanc Widnes', as Hardie (1950) called it. Whatever the horrors of its urban landscape, its attractions for the trade were such that it grew rapidly and sometimes at the expense of other areas.

Owing to changes in ownership and the quick failure of some enterprises, it is difficult to be sure exactly how many Leblanc works were built in Widnes, but there were at least twenty-two, of which the last was put up in 1884.

Between 1852 and 1861 the alkali trade of south Lancashire generally tripled; much of this increase must have been in Widnes, though it is impossible to say just how much. (*C.N.*, 21 Sept. 1861, p.156.) The American Civil War constituted a considerable blow and caused a lull in the town's headlong growth, no new works being built between 1859 and 1865. But with the ending of the war, expectations in British alkali in general were high—at one works on Deeside built in 1863, the promoters expected a profit of £1 a ton (*M.J.*, 14 Jan. 1863, p.81). Expansion then swelled into a boom up to the Franco-Prussian War. It was helped by changes in business organization. Limited liability concerns were now rapidly emerging. Provision for joint-stock companies and limited liability had come in a series of Acts from 1844. A great codifying Act (the Companies Act 1862) eased their further development. The outstanding period of expansion in Widnes was the ten years from 1865, which indicates that the concessions made by the London and North Western Railway in 1864 certainly did not restore the attractions of St. Helens and suggests that the modest resumption of construction there might have been mainly due to the general boom in the alkali trade. In the fifteen years following 1863, Widnes capacity increased fourfold. (Noxious Vapours, 1878, p.5; Fig.13.)

Table 30. *The establishment of Widnes alkali works: time distribution*

1850–4	4	1870–4	8
1855–9	1	1875–9	–
1860–4	–	1880–	1
1865–9	8		

Based on: Hardie, 1950; Widnes, 1872.

The canal works

The works of the canal area of Widnes were all established in the fifteen years to 1865 (Table 30). John McLellan, who followed Hutchinson in applying for a lease in the summer of 1848, chose a location along the St. Helens railway, but for almost all the other early arrivals the canal dominated the choice of site. William Gossage chose Woodend in 1850, by the canal but with sidings on the Runcorn Gap railway giving direct access to Wigan and to St. Helens coal. Five years later he closed his alkali works and built a soap-works on the same site. (Noxious Vapours, 1862, p.83.) Most of the rest of the canal-side development was undertaken by entrepreneurs from other inland centres. In February 1852 the Muspratt family applied for a lease of land, and eventually received approval for a site a little further up the canal from Hutchinson's. The next year Hutchinson acquired more land and seven years later built a second smaller works on Waterloo Road. Hutchinson's manager Henry Deacon, who had come from the glassworks of St. Helens, left his employment in 1853 and with the co-operation of Pilkingtons, his former employers, built his own works. (Allen, 1907, pp.164, 165; Ashmore, 1969, p.134.) The Pilkington interest in Deacon's works was given up in 1855, but ten years later

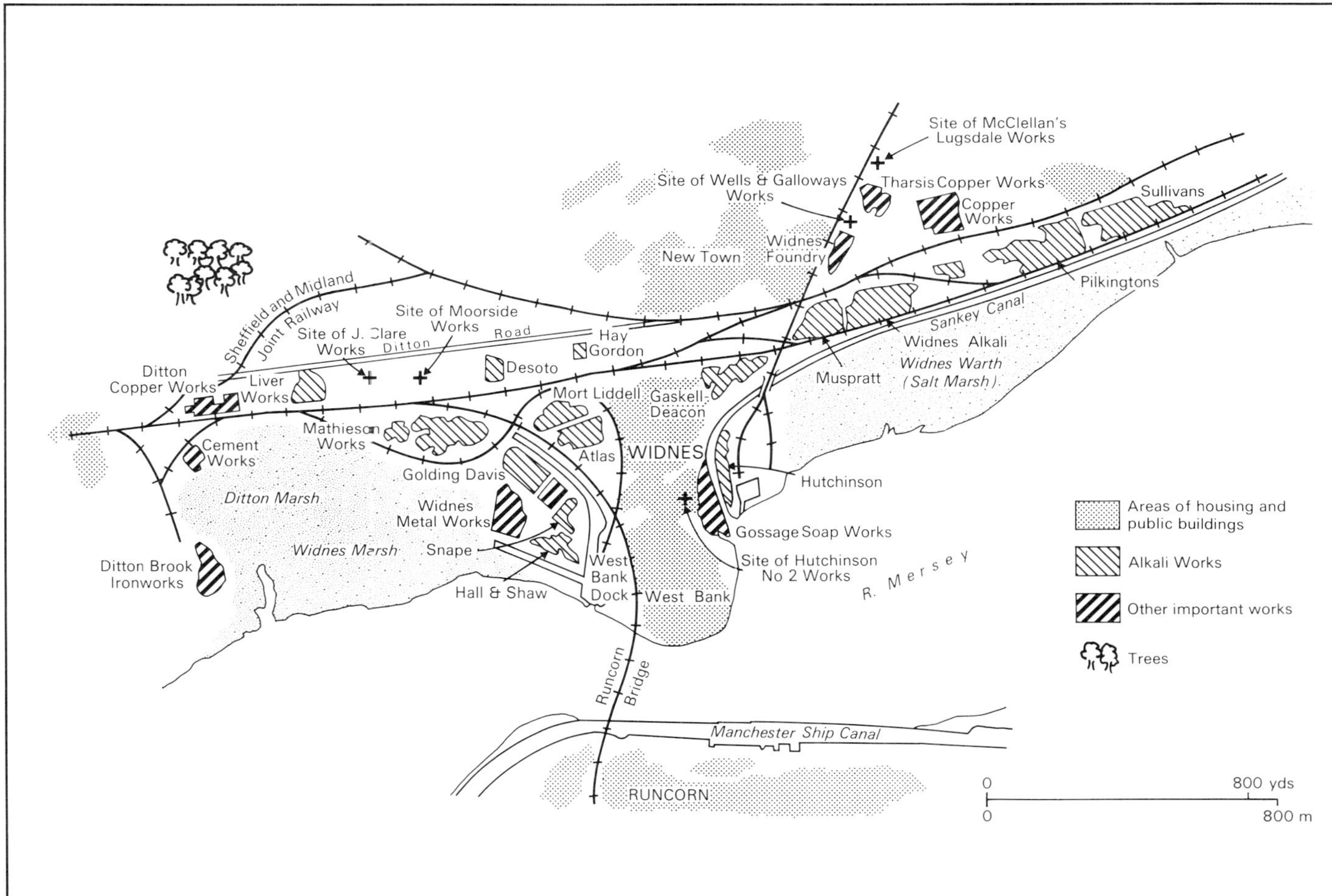

Fig. 13. The Widnes district in the early 1890s Based on Ordnance Survey, Six Inch Map, revised 1894/5

William Pilkington established the Mersey Chemical Works. The Widnes Alkali Works was put up in the same year. In 1867 chemical development on the canal banks came to an end when four St. Helens men—Sullivan, Sievwright, Crossley, and Brock—built the British Alkali Works. By 1870 well over 1,500 men worked in the eight alkali and soap works along the Sankey Canal. (Hardie, 1950, *passim.*) Attention now shifted to the so-called Marsh district and to Ditton Road.

Widnes Marsh

John Hutchinson had bought up land in Widnes. One of his acquisitions was the Widnes Marsh on which he leased the right to dump waste. Within ten years this was so successful that the land was mortgaged for £20,000 and Hutchinson prepared to develop it with its own dock and railways in the hope of attracting new industries. In 1865 he died with these plans uncompleted (*Weekly News Widnes*, 20 Aug. 1965), but as the last works were being built along the canal the first were going up on the Marsh, beginning in 1866/7 with works built by Hutchinson's current second-in-command. Four other alkali works and a plant to extract copper were established in the Marsh region between 1866 and 1873. The most important and most interesting of these was the works built by the Scot Neil Mathieson, formerly a manager with Johnsons of Runcorn but who now co-operated with two Widnes soap firms, Gossage and T. S. Timmins. Mathieson's became one of the most important works in the town. Two other works were built between the viaduct approach to the Runcorn Bridge and the town centre.

By the end of the period of Marsh development a start had already been made on the last area of Widnes to be opened for alkali, Ditton Road, with access to rail on the south side of the works. Between 1870 and 1874 four works were built along Ditton Road. One, the Desota Chemical Company, was an almost wholly Liverpool organization in terms of shareholders. Ten years later, when competition was becoming keen in the alkali trade, the final Widnes alkali establishment emerged. An alkali plant had been built in Lightbody Street Liverpool on the banks of the Leeds and Liverpool Canal in 1867, in the same general locality as Muspratts had chosen forty years earlier. In 1884 the firm was wound up and the Liver Alkali Company was formed to build a new Leblanc works on Ditton Road, the most westerly of all the Widnes operations. By this time other works in the town were being closed. The era of rapid growth was at an end.

In spite of the later development of both the Marsh and the Ditton Road areas, the biggest works were generally along the canal, though Mathiesons was an exception. In 1892 Mathiesons employed 438 men; four other Marsh works had an average of 149 persons. The six surviving canal works employed an average of 577 men each. (Hardie, 1950, *passim*; Table 31.)

The canal works also proved to have the highest survival rate. After 1920 the only works still in operation elsewhere were the Mathieson and the Golding

Table 31. *Widnes works, 1891*

Area	*Number*	*Capacity all products* (thousand tons)
Canal	5	381
Marsh	3	122
Ditton Road	1	12
Viaduct	2	25

Source: U.A.C., 1896b.

Davis plants. Along the canal the Gaskell-Deacon, Muspratt, Pilkington, Sullivan, and Widnes alkali works all survived to pass into new combinations of works. Hutchinson's works was closed in 1919, seventy-one years after the first application for a lease had begun the industrialization of Widnes. (U.A.C., *c.* 1930.)

Comparative costs

Lancashire made rapid headway at the expense of the North-East in the 1870s and 1880s (Fig. 14). Even so, there is evidence that costs of production remained very close. Richardson and Watts (1863, pp. 317–18), on comparing costs of raw materials in soda ash production in South Lancashire and at the

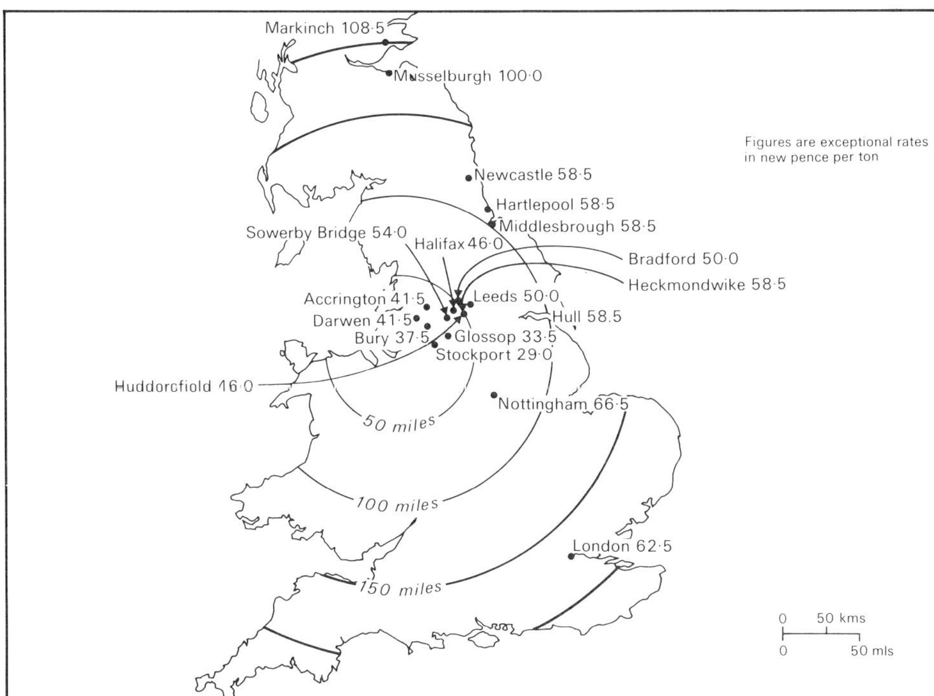

Fig. 14. Railway freight rates from Widnes on alkalis 1880/1 *Source:* S.C. on Railways, *B.P.P.*, 1881, XIII

Jarrow Chemical Works, suggested figures per ton of £4. 8*s*. 0*d*. and £4. 9*s*. 9¾*d*. respectively. Mond reckoned that Widnes works could manufacture ash for £8. 19*s*. 10*d*. and deliver it in Liverpool for an over-all cost of £9. 1*s*. 10*d*. He added, 'At the best situated works in Newcastle the probable cost is £9. 11*s*. 0*d*.' (Mond, 1872). In the same year he made another estimate which narrowed the gap a little—£9. 7*s*. 0*d*. for Widnes soda ash delivered in Liverpool and £9. 15*s*. 0*d*. at 'the best situated works in Newcastle' (Dick, 1973, p.116). In the ensuing years Brunner Mond & Co. kept a close watch on the Leblanc industry and its costs. Their cost ledgers contain a detailed assessment for 1884 of the comparative costs at Widnes, Runcorn, and Ditton, the outcome of which was to show that Widnes was 1¼*d*. cheaper than Ditton, but 1*s*. 5¼*d*. a ton lower in over-all production costs than Runcon (Mond, 1884). In November 1888 they made a close study of Newcastle and Widnes which seemed to show that Newcastle was cheaper by 10*d*. a ton in soda ash (Table 32). A full commentary on these figures, however, pointed to an increase on the Tyneside costs and a decrease in Widnes. As evidence of the care taken in cost assessment and of the indeterminacy that still remained the comments are quoted in full.

Table 32. *Costs of 58 per cent soda ash in Newcastle and Widnes, November 1888*

Item	*Weight* cwt qrs lb	Newcastle *Unit cost*	Newcastle *Cost per ton of soda ash* s. d.	Widnes *Unit cost*	Widnes *Cost per ton of soda ash* s. d.
Pyrites	16 0 14	12*s*. per ton	9 8	13*s*. 6*d*. per ton	10 10½
Salt	29 0 4	9*s*. per ton	12 11	9*s*. 6*d*. per ton	13 8
Nitrate	— — 24	9*s*. 3*d*. per cwt	2 0	9*s*. 3*d*. per cwt	2 0
Limestone	31 0 15	3*s*. 9*d*. per ton	5 10	4*s*. 5*d*. per ton	6 10
Saltcake slack	24 0 21	4*s*. per ton	4 10	4*s*. per ton	4 10
Mixing slack	12 1 23	5*s*. per ton	3 1¼	4*s*. 6*d*. per ton	2 9¾
Firing slack	38 1 21	4*s*. per ton	7 8¼	4*s*. per ton	7 8¼
Saltcake wages			10 1		10 1
Soda ash wages			13 10¾		13 10¾
Waste removal			7		2 1
Maintenance and stores			11 10		8 6
General expenses			10 9		10 9
Cost net at the works			£4 13 2¼		£4 14 0¼

Source: I.C.I., Runcorn Records.

The figures for pyrites and salt will require very close working. I think the figure for nitrate—24lb—too low. This would be equal to 15lb. only per ton of saltcake, and this is certainly much below the average used in Widnes, though it may possibly be correct on the Tyne, where they have been accustomed to work with more chamber space than we, but if it is intended to include saltcake made by Hargreaves process in Widnes (which requires no nitrate) will be probably about correct, but still low. And is the price on the Tyne the same as in Widnes? I should have expected to see it higher.

Limestone is very low in weight, and I think there must be an error in the Newcastle price. When last I visited works on the Tyne they crushed their own stone, probably they do so now, so the cost ought to be 3*s*. 9*d*. plus 6*d*. say 4*s*. 3*d*. per ton. The Widnes price should be 1*d*. less say 4*s*. 5*d*. Surely the saltcake slack is too high. In Widnes 11cwt. per ton of saltcake is the figure, which equals 17½cwt. per ton of 58 percent ash. On the other hand I think the mixing slack should be a little higher—say 13 cwt.—and the total less—say 70 cwt. per ton in Widnes. But is it correct to take the same weights as used in both districts. If slack in Widnes is of better quality as generally agreed, surely the consumption ought to be less than at Newcastle.

Wages. Are they the same per ton in both districts? The Tyne men used to claim to pay less than we did. Waste Removal. I doubt the Tyne figure. 7*d*. may be all that is paid for boat work, but it must cost something for putting the waste on board the boats over and above what is paid for discharging vats, and this should be added to the cost of removal, or else to wages. In Widnes I think on further enquiry the cost may be reduced to 2*s*. 1*d*. I find there are some makers who are removing their own waste at a cost of 6*d*. per ton of waste, say 1*s*. 7½*d*. per ton of 58 percent. Total cost on the Tyne. Mr. Mond told me in October last Allhusen said his cost was £4 per ton of 48 percent f.o.b., that is of course packed. The above figures work out to £3. 17. 1½*d*. for 48 percent.

It is not known who is the author of these estimates but obviously he was extremely well informed (Table 33).

Table 33. *Costs of soda ash production in Widnes and Newcastle, 1872 and 1888*

	Widnes(cost delivered at Liverpool) £ *s*. *d*.	*Newcastle* £ *s*. *d*.
Mond 1872	9 1 10	9 11 0[a]
Brunner 1872	9 7 6	9 15 0[a]
B/Mond 1888	4 14 0¼	4 13 2¼

[a] Cost 'at best situated works on the Tyne'.

Source: I.C.I., Runcorn Records.

In 1852 the North-East had decomposed nearly 50 per cent more salt than the North-West. By the mid-1860s the North-West was already ahead, though it is not clear by how much, for the boundaries of the districts for which figures were provided had been changed. By the mid-1880s Lancashire was over three times as important as the North-East and maintained this lead into the nineties (Table 34).

Table 34. *Tonnages of salt decomposed in the North-East and Lancastria 1852–1892*

	1852	*1866*	*1887*	*1892*
North-East	57.9	151.7	117.2	93.2
Lancastria	40.1		380.0[a]	275.5[a]

[a] Excludes east Lancashire, but includes North Wales.

Sources: various.

CHAPTER 9

The Alkali Trade and the Transformation of South Lancashire

The alkali trade of the Mersey area took shape in an already rapidly industrializing and urbanizing region. In 1845, with remarkable perception, a Board of Trade committee recognized a wide urban/industrial complex occupying most of south-east Lancashire:

> Within a circle of 15 or 20 miles around Manchester, a population of upwards of a million and a half are concentrated, who are almost without exception either actively engaged in, or directly dependent upon the great staple manufacture of cotton. The whole of this district may, in fact, be considered as one vast workshop, where production is constantly going forward on a scale hitherto unparalleled in the history of human industry. (Board of Trade, 1845, p.361.)

Wigan and Warrington lay just within the outer limit of that industrial region. One of the major contributions of the alkali industry to the population and settlement geography of Lancastria was to build up three other nuclei just beyond, thus extending the edge of the 'Manchester region' and effectively making the link with the more concentrated area of population round the Mersey estuary.

In the 1830s Runcorn, St. Helens, and the little village of Widnes had a combined population of some 12,000; by the end of the nineteenth century they contributed to the wider south Lancashire complex which, with the new axis along the Manchester Ship Canal, created a great growth area for some 42 miles from the mouth of the Mersey to the foot of the Pennines beyond Oldham and Stalybridge, a growth area which, in the eyes of enthusiastic North-Westerners bid fair to eclipse London and the South-East.

The Chemical towns of south Lancashire

Chemical manufacture affected even the character of the biggest units of south Lancashire. Before Muspratts' works were built, Everton was 'the court end of Liverpool'. Later the fumes were carried up to Everton and destroyed the vegetation; as a result, 'the more respectable inhabitants left the district; the property was very much deteriorated, and at the present moment most of the good residences are being pulled down and streets of small houses are being built upon the land'; in addition a 'residential prejudice was created against the district' (Noxious Vapours, 1862, pp.114–25). Runcorn doubled in population from 1861 to 1891, and altered in character—as late as 1855 it had been described as 'a favourite resort of sunbathing visitors in the summer' (Knight, 1855). In the nineties recession in chemicals brought severe distress to the town and by 1897 800 houses were empty and rateable values had slipped sharply. During the period 1881–91 there was an increase of 8,000 in the population of the Runcorn registration district of which 2,000 was due to net in-migration. In the next inter-censal period natural increase remained at 6,000 but there was an over-all decline of 4,000 in population as a result of a net out-migration of 10,000 persons. From 1901 to 1911 growth in Runcorn population was resumed but at a slower rate than in the eighties. (Lawton, 1953, pp.266–7; *C.T.J.*, 18 Dec. 1897, p.391.)

St. Helens was a county borough with a population of 84,000 by 1901, though within living memory it had been little more than a village. In terms of occupations it was the most diverse of the chemical towns: by the mid-nineteenth century it had glassworks and coalworks, copperworks derived from those which had smelted Parys copper late in the eighteenth century, potteries, iron and brass foundries, and a brewery as well as the chemical works. The area was also well known for the production of excellent watch movements. In the nineties it retained most of these trades and was a market centre as well. By now, however, the chemical trade in St. Helens was shrinking and the town was dominated by glass-making. Pilkingtons alone employed 4,000 by the mid-nineties, and among the other glassworks around the town Cannington Shaw claimed to be the largest glass bottle manufacture in the world. Even so, St. Helens played a significant role at this time in the North-West's pre-eminence in the manufacture of chemical plant (Fig.15).

In 1841 Widnes and district had a population a little over 2,000. The main occupations were farming, the weaving of canvas for sails, wire-drawing, tool-making, file-cutting and watch-making rural extensions of trades typical of Warrington and Prescot. It was, as John Hutchinson later recalled, 'quite a rural district' when he arrived (Noxious Vapours, 1862, p.161). The boom of the chemical trade made it much more of a one-industry town than St. Helens. By 1855 Widnes township had five chemical works, a manure works, one oil manufacturer, three sailcloth-makers, four file-makers, nine wheelwrights, joiners, blacksmiths, and iron-workers, and thirty-three farmers. (Mannex, 1855, pp.810–12.) By 1872 of thirty-five industrial plants in the town, twenty-four were chemical works. Eight years later 50 per cent of the industrial population of Widnes were chemical workers. (Widnes, 1872; Lawton, 1953, p.266.) The rateable value of Widnes increased with great rapidity (Table 35).

Table 35. *Rateable value of Widnes and Ditton* (£)

Year	Value
1841	6,236
1851	8,134
1861	21,846
1875	86,114

Source: Noxious Vapours, 1862 and 1878.

The danger of concentrating on chemicals was appreciated, and attempts were made to diversify the structure. One example of a directly associated diversification involved the establishment of three copperworks in the town.

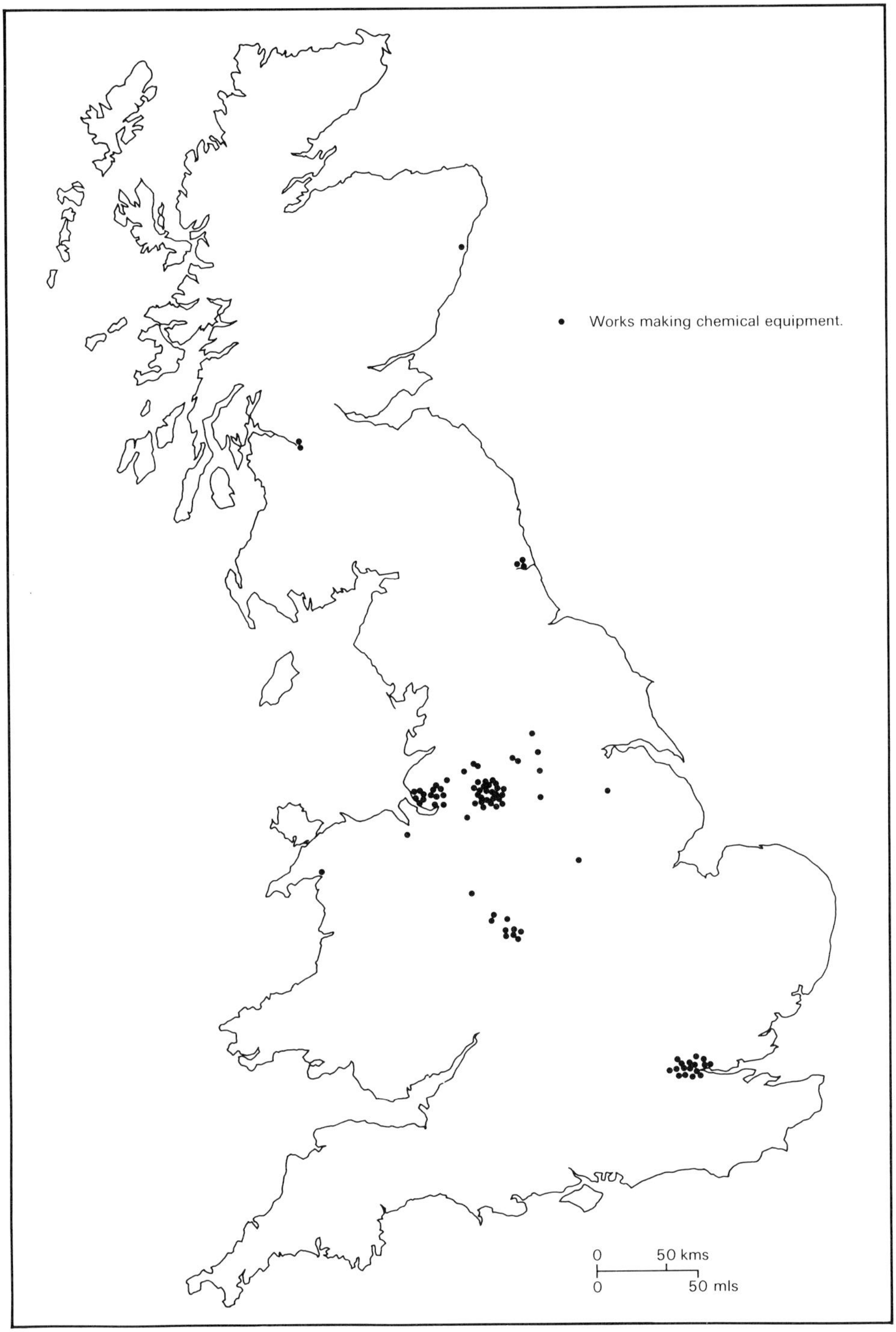

Fig. 15. British manufacturers of chemical plant and equipment 1892/3 Based on *C.T.J.*, 1892/3

The Ditton Brook Iron Works in the 1870s made pig-iron and plant for chemical works. (Worrall, 1876.) The Widnes Foundry Company not only made a speciality of chemical plant but around the turn of the century, as leisure and holiday-making increased, was particularly prominent in promenade and pier construction. New firms came in as early ones declined. Thus the site of the Desota Chemical Company in Ditton Road was bought shortly after the turn of the century and developed by the Birmingham Corrugated Iron Company in its search for sites on which nuisance-creating activities would be tolerated and which afforded better access to the ports of shipment overseas than sites in the West Midlands.

As a result of attempts at diversification Widnes became to some extent an iron and non-ferrous metal processing town and an engineering centre, but it remained dominated by chemicals. In 1893 the United Alkali Company employed almost 9,000 people there, out of a total population of 30,000. (*C.T.J.*, 22 July 1893, p.53.) With the obvious difficulties of the Leblanc industry in the nineties, interest in broadening the economic base grew. A committee was appointed to study this problem and counteract the 'very general impression abroad that Widnes is fit for nothing but chemicals, soap, copper and iron works'. It saw that there was a special need for employment for women and girls and suggested the establishment of clothing and boot and shoe manufacture. Emphasis was laid on the town's good transport facilities, low rates, its pure and abundant water and very cheap gas, but it was difficult to promote a better image of Widnes. 'That Widnes possesses great advantages as a manufacturing centre, none who are acquainted with the place will deny, but at the same time it is difficult to force the growth of a town in an old country like this, where trade has settled into certain grooves and localities' (*C.T.J.*, 3 Apr. 1897, p.234).

Urban fabric and social structure

By the 1850s St. Helens was a 'straggling ill-built place' (*English Cyclopaedia*, 1855). Incorporated in 1858 it became a county borough twenty years later. In 1873 it had, in the Lancashire tradition, a town hall which was 'a fine building in the Gothic style', and by the nineties had a 'great number of new streets and has undergone considerable improvements'. It had two parks and a number of public buildings, including the theatre and music hall. Even so, to a local writer it seemed that 'no town in England of similar size was so deficient in public provision' and he referred to its 'pre-Adamite condition . . . as regards public institutions, whether literary, artistic, scientific, sanitary or social' (*St. Helens Lantern*, 1887; *Comprehensive Gazetteer*, 1895).

Widnes started from a lower level. Its population increased some fourteenfold in less than fifty years. Attempts were made to improve what the *Daily News* in 1888 described as 'the dirtiest, ugliest and most depressing town in England' (Morgan, 1972, p.4). At the inquiry held by the Commissioner of the Privy Council into its petition for incorporation in 1891 Widnes's facilities were listed. There were board schools for 3,600, five denominational schools, and two non-sectarian ones. Three political clubs operated in the town; it had a free library, town hall, theatre and drill hall, a market, six post offices, six passenger and five goods stations. (*C.T.J.*, 4 Apr. 1891, p.219.) In 1894–6 a new free library and technical school was built at a cost of some £10,000, designed to give the town respectability as the late Victorians saw it 'with a very handsome front to the square . . . and in form and colour the schools will come in admirably with the town hall and other new buildings roundabout. The style is English Renaissance, and the material to be used is red Ruabon brick, with terracotta dressings' (*C.T.J.*, 28 July 1894, p.51). However, the builders of the new Widnes were struggling with over forty years of speculative boom-town building. As Councillor Lewis put it a few years later, 'The "plan" of the town is a striking example of how not to do it' (Lewis, 1911a). Lewis filled out what he had in mind, 'As a matter of fact it has grown up irregularly in an elongated fashion along the old highway, hugging the line of works from north to south for about two miles, bulging out here and there as necessity required and opportunity offered'. He concluded wearily, 'Under the circumstances it is difficult to see how it could have been otherwise' (Lewis, 1911b).

Within these ill-planned, drab if improving towns, the growing and uncertain population had a skewed social structure. The towns were definitely working class: 80 per cent of the jobs in Leblanc works were labouring jobs. (Millward, 1955, p.96.) Generally the larger the town, the wider the industrial structure and the socio-economic class. Age of settlement was also probably of importance, the older, if smaller, units having a more diversified society than the booming newcomer. In all respects, of the three main chemical towns, St. Helens was the most, and Widnes the least, favoured. Yet even in St. Helens the low figure for professional occupations, and, perhaps even more suggestive, the low proportion of the population in domestic service emphasize the imbalance of its employment and socio-economic structure. If data were available, Widnes would undoubtedly provide an even more extreme case (Table 36).

All chemical towns had narrow societies, though there was division of opinion between those who viewed their socio-economic condition from different perspectives—those who made money there, those who regretted the character of these towns but saw it as a necessity, and those who roundly condemned it. The uncertainty of employment, the horror of the work which had to be done, and the poor social provision were critical aspects.

It is perhaps not surprising that jobs were created or abolished with little concern for the workers who drifted in and out of the communities. In the year to July 1898, the population of Widnes fell by 568 or 1.8 per cent; 582 houses were uninhabited at the end of July 1898, 9.8 per cent of the housing stock. 'It is', reported the Medical Officer of Health, 'most difficult to give a correct

Table 36. *Professional, domestic, and commercial occupations in proportion to total population, 1901*

	Population (thousands)	*Professional* (including Civil Service and defence) %	*Domestic* %	*Commercial and conveyance* %
England and Wales	32527	2.9	6.1	6.9
Liverpool	685	2.5	5.5	14.1
Manchester	544	2.1	4.0	9.4
Bolton	168	1.6	2.9	5.2
Oldham	137	1.4	2.4	4.9
St. Helens	84	1.5	2.7	4.9

Note: The low rank of Oldham is probably in part related to proximity to Manchester.

Based on: Census of 1901.

estimate of the population in a township so largely composed of the working classes. Due to the closing of works, etc., many workmen, in search of employment, migrate from one town to another, taking their wives and families with them, consequently many houses become empty.' (Widnes, 1899, p.4.)

Working conditions were always poor and sometimes appalling. Some tried to put a brave face on it. In September 1894 A. E. Fletcher, Chief Inspector under the Alkali Act, gave a lecture in University College Liverpool to the Congress of the Sanitary Institute (*C.T.J.*, 6 Oct. 1894, p.209):

> The general mortality too, of Widnes compared favourably with that of other manufacturing towns in Lancashire. It would appear that the workers in chemical factories, though exposed to sights sounds and smells that might alarm the stranger, have not more of their share of the toil and strain that must ever accompany honest work. The workers, doubtless, preferred their employment to the irksome confinement of the city office or the shop counter and learned to take a wholesome pride in the productive character of their labour.

Evidence was in fact already substantial that this was a far too rosy view of the reality.

An account of St. Helens in 1846 noted that when working men 'by any chance inhale more than the usual quantity of gas, vomiting and fainting are brought on and they are obliged to be carried out of the works for air'. Escaping gas burned clothes. (Barker & Harris, 1954, p.283.) In the eighties and nineties every issue of the technical and trade paper the *Chemical Trade Journal* contained reports of violent death in the chemical works, but between late 1889 and mid-1891 this journal also reported a most disturbing correspondence on normal conditions in the works. A chemical worker, engaged in the job of firing the caustic soda pots, wrote to Gladstone about conditions of work:

> For these men there is no New Year's day, no Christmas Day, no Good Friday. These works ignore all such days. *It is one eternal round of drudgery.* I have been twelve years in my present place, and during that time I have not had the pleasure of enjoying one single New Year's day, Christmas Day or Good Friday in the midst of my family . . . .
>
> It may be argued that we have all the evenings of the week we are on the day turn. Yes, but then, it requires all these evenings to resuscitate our enfeebled energies.

Gladstone replied with sympathy to this moving letter but showed both how ignorant he had been of industrial conditions no more than seventeen miles from Hawarden Castle and how ineffective even the concern of a Prime Minister could then be. 'I have been sadly and painfully impressed by the facts set out so clearly in your letter, and I shall keep it by me for further reflection. I did not know that in trades of such a class such cases were possible.' (*C.T.J.*, 30 Nov. 1889, pp.333, 334; 6 June 1891, p.363.) At about this time P. J. King, President of the newly formed Chemical and Copper Workers Union, visited the workhouse at Whiston some three miles from St. Helens and five from Widnes. As recorded in the union's first Annual Report,

> he found that every fourth man was formerly a chemical worker or a chemical labourer in the union, and that they had been driven there by ill health or poverty, or more frequently both, in fact the institution was a refuge for the poverty-stricken, used by employees of our wealthy chemical manufacturers. Many of the men detail their struggles for an existence, showing how the unfavourable conditions under which they had been compelled to work, thus doom them to a pauper's life and a pauper's grave. (*C.T.J.*, 6 June 1891, p.363.)

Men died in Whiston, not in the works, for they were dismissed before they reached that stage of decay. (Sherard, 1897; Blatchford, 1899.) The transactions of the Manchester Statistical Society (1899, pp.138–40) provided a dignified but frightening enough account of the relation between alkali works and health: 'The occupation [of soda ash manufacture] is not only very irritating to the eyes, skin and lungs, but it is very exhausting, and the workmen are in consequence addicted to the excessive use of alcohol. Gasing with chlorine occasionally occurs. The teeth are destroyed and appetite fails.' The sufferings of the workers did not always meet with a sympathetic response from those in more favourable positions: 'While we admit that all the detailed instances of death or premature decrepitude so vividly depicted may be perfectly authentic, still there is a medium in all things' (*C.T.J.*, 4 Mar. 1893, p.141).

What was an extremely hard and degrading situation for the workers meant profit for landowners and managers. The landowners made their money out of mushroom growth, and lived well away from the appalling visual effects which it sometimes had. About a quarter of a century after Widnes development began, John Bibby of Liberpool, who owned 210 acres of what had been good farm land on the north side of the London and North Western Railway, now occupied in part by the Sullivan and Pilkington works, noted, 'I have never had experience of a like increase in prosperity' (Widnes, 1872, p.4). Most managers soon took up residence away from the chemical towns. A local paper summed up a serious problem: 'Our leading traders seem to have an eye to only one thing—that is to make all the money possible, and the quicker the better, and then off from the "Inferno" with "an-Dante" speed" (*St. Helens Lantern*, 1887, p.57). St. Helens was fortunate in at least one of its manufacturing families, the Gambles, who concerned themselves with local government and with social improvements.

In Widnes the entrepreneurs Gossage, Hutchinson, and Deacon lived fairly near their factories, but by the early 1860s Mond found that most officers of the chemical companies lived at Appleton. Ferdinand Hurter lived at Wilmere House in open country north of the town in the eighties and travelled to work by pony trap. (Hardie, 1950, pp.22, 90, 160.) Soon there were complaints that the wealthy of this town too took train after work to homes well away (*Minstrel*, 1906). Eulogizing the men of business who had pioneered the Widnes chemical trade, Allen (1907) none the less saw the distinction between private and social gain. Perhaps no Lancashire alkali manufacturer epitomizes the self-interested capitalist as well as Christian Allhusen of Gateshead who died in 1890 leaving a personal estate of £1.1 million. Allen (1907) wrote of him with a scathing undertone thoughout—'he reached the haven of opulence and commanding social status'—and ended with a damning epitaph 'Born in 1806, making money while he was in his teens, he was favoured with a career of exceptional length and opportunity and in his eighty-fourth year he died a millionaire' (p.248). Allhusen, according to Allen, left no permanent benefaction to Tyneside, 'such expressions of a desire to benefit the people do not appear to have been consonant with his views, and he is not singular in the opinion, that the man who, in pursuit of his own personal enrichment and who with broad and intelligent conceptions of the science of business seeks his own interests, is a benefactor to the people' (p.245). (It must be stressed that other accounts of Allhusen are more favourable.) Allen went on with the conviction, so dear to *laissez-faire* economists, and so much at variance with the character of the communities their business created that the entrepreneur's self-interest necessarily benefited all—'his business becomes a nucleus whence other trades and occupations spring, and a fine works erected on a site that, by its suitability, attracts other similar works, gives a character to a population, and tends to promote the institutions that elevate and enrich social life' (p.245). John Hutchinson gave every indication that, had he lived, he might have been a Lancashire Allhusen. Yet, while fortunes were made out of chemicals, the viability of the Leblanc industry was slipping; the working men were labouring under terrible conditions and their towns were becoming notorious for their repellent landscapes.

The alkali trade and the environment

Matthew Arnold labelled St. Helens 'a hell hole'; Ruskin seems to have been particularly appalled by Widnes. No one pretended that either was attractive, but many commentators argued that jobs and expanding populations made the affront to amenity justifiable or that wider benefits to society justified the sacrifice of a few localities. The Consular Agent of the United States, when asked what he thought of St. Helens in the 1880s, observed, 'it may be smoky and is certainly not picturesque, but it is playing a useful part in the all important economy of modern commerce.' (*St. Helens Lantern*, 1887, p.111).

Some twenty-five years earlier it had been reported that everything in the town exposed to chemical vapours had become slaty blue and that shopkeepers therefore painted their shutters that colour (*C.N.*, 18 Oct. 1862, p.202). The *Chemical News*'s review of the *Report of the Select Committee of the House of Lords on Injury from Noxious Vapours* contains a comprehensive and depressing account of the effects of pollution on the local environment. Blatchford's description (1899, p.15) was more colourful: 'St. Helens is a sordid ugly town. The sky is a low-hanging roof of smeary smoke. The atmosphere is a blend of railway tunnel, hospital ward, gas works and open sewer. The "features" of the place are chimneys, furnaces, steam jets, smoke clouds, and coal mines. The products are pills, coals, glass, chemicals, cripples, millionaires and paupers.' Yet by the 1890s St. Helens had a redeeming feature in some carefully tended public parks (Bannerman, 1897, pp.36–7).

Allen (1907, p.23) noted that a few years after development began, in Widnes, apart from the sprawling streets there were 'the works that were belching forth volumes of most deleterious gases and clouds of black smoke from chimneys of inadequate height, with trees that stood leafless in June, and hedgerows that were shriveled in May. The air reeked with gases offensive to the sight and smell, and large heaps of stinking refuse began to accumulate.' Gaunt skeletons of trees and hedges still stood in the early 1880s (Grindon, 1892, p.144). In his early days in Widnes, Gossage lived between two chemical works. He became very ill and attributed this to the gases. On his doctor's advice he moved away, but did not discuss what might happen to workers who could not move. (Noxious Vapours, 1862, p.89.)

In 1888 Hurter gave a lecture entitled 'The Philosopher's Stone' (Hurter, 1888). He first sketched a rural Eden from which Widnes's development had begun: 'Once upon a time there was a small village situated near the shores of a beautiful large river in a lovely country. The peaceful inhabitants of that village, most of them tillers of the soil, enjoyed themselves in the simple fashions of those times.' That year the contemporary Mersey scene was described (Fisher Unwin, 1888, pp.18–19): 'It is permissible to imagine that the ashes of Sodom and Gomorrah immediately after the destruction of those wicked cities emitted no more pestiferous odours, or were more devoid over all their wide area of one single green leaf or sign of vegetation, than those twin towns of the Mersey—Widnes and Runcorn.' But even this writer could not put aside his Victorian values: 'The latter have, however, this advantage, they are by no means *dead* cities of the plain. Great and busy industries are carried on day and night in both, and in each large fortunes have been and are still being made . . .'. Alkali provided not only the personal income but also the chief source of municipal income in these communities. By 1912 the valuation of Widnes had reached £173,000. Some two years before, the gross rateable value of the United Alkali Company works in the town was £50,000. (Whitaker, 1913; Borough Guide, *c.* 1910, p.13.)

The environmental impact of the alkali industry may be considered in terms of the buildings and equipment of the trade, the effect of gases and liquid effluent on the landscape, and the solid wastes of the works. In the late 1890s the industrial landscape of the chemical towns was summed up in a local survey (Bannerman, 1897, pp.38, 39):

> The predominating features of these works are wide-spreading gloomy sheds, more or less open, and disclosing in the ground spaces great vessels and kilns, with gleaming fires here and there when furnace doors are opened; overhead are vast leaden chambers, supported by strong wooden frameworks, and among them rising still higher are towers more dark and mysterious than that to which Childe Roland came, while to still loftier regions are lifted the tall chimney stacks.

The problem of gases and smoke from the alkali works was a continuing one. The Alkali Act of 1863 set stringent standards; some of the alkali-makers seemed almost surprised at the ease with which they met them. One large producer, who had averaged £150 a year damages for the preceding four years, installed recovery equipment for £300. In the summer of 1864 some of the fruit trees which had not blossomed for many years did so again and roses grew where none would grow before in the areas near the alkali works (*The Times*, 23 May 1865, p.13). However, in the expansion of the early seventies, many new alkali works or extensions were not provided with adequate control over the emission of gas, and new and more stringent controls had to be established in the new Alkali Act of 1874.

In the 1860s an expert valuer of damages from noxious vapours, Rothwell, noted how grass exposed to the gases in winter could in spring be pulled up dead in handfuls. Sometimes an area over a breadth of a mile and up to two miles or more from the works was covered by vapour which affected both eyes and breathing. All the land within a quarter of a mile from works at Weston and towards Halton from the Runcorn works was useless as a result of the gases and had been acquired by the manufacturers (Smith, 1867, p.144). Allen denied that destruction of trees in Dunham Park fifteen miles away from Widnes could be blamed on the alkali works, attributing it instead to shallow sandy soil, drainage problems, bleak aspect, and hard winters. The chemical manufacturer often had to pay for damage to the celery crop around Widnes when the real cause was caterpillar infestation. It was common to make the 'chemics' the cause of diverse farming ills (Allen, 1881, p.342). Weldon, who in his later years visited many European chemical works, noted 'In all countries to which my experience extends I have noticed that farmers have a tendency to prefer to exploit any chemical works in their neighbourhood rather than to exploit their land' (*J.S.C.I.*, 1882, p.42).

Smoke was an additional nuisance. Widnes in 1887 used 960,000 tons of coal, the combustion of which released into the air an estimated 120 tons of oil of vitriol daily (*C.T.J.*, 31 Mar. 1888, p.203). The ugly sulphurous waste from the black-ash furnaces caused drainage problems. The Sankey Brook, whose valley the canal followed down to Warrington, became a chemical nightmare

(*C.N.*, 2 Oct. 1868, p.160). In pre-industrial days Bowers Brook, which discharged near Widnes dock, had been a trout stream. It was later culverted and became a channel for chemical waste. Tank waste itself created a great nuisance, in loss of land, in the form of an offensive landscape, in the spontaneous combustion of sulphur in the heaps when the weather was dry or the run-off of sulphuretted water in wet weather. In one case before 1860, in the St. Helens' area, £1,000 an acre had to be paid for land on which to deposit this 'galligu'—in depths of up to 15 or 16 feet (Noxious Vaœours, 1862, p.25). Chance's recovery process promised some relief, but it did not deal with the greater part of the waste, and not all could afford to invest in it. Some suggested that the waste should be turned into slurry and be disposed of in coal-mines. In the fifties Deacon used it for making floors. Some was employed for building blocks, and some for footpaths, but, even so, in the early 1860s these outlets absorbed not more than 2 per cent of the production. (Noxious Vapours, 1862, pp.88–9.) Over thirty years later, efforts were made to interest farmers in its use as a manure: a top dressing was left on the soil to get rid of excess sulphur and then the waste was ploughed in to cleanse the ground by killing slugs and weed seeds. Some years, thousands of tons were carted away by farmers from Widnes works—the railway companies were reluctant to handle it (*St. Helens Lantern*, 19 Apr. 1888, p.109; *C.N.*, 21 Sept. 1861, p.156; *C.T.J.*, 26 Aug. 1893, p.128). Waste continued to accumulate nevertheless.

Waste did not become a problem around Widnes for about fifteen years after the foundation of the first works, being used in the mean time to reclaim and level ground. After 1865 tips began to appear in the Marsh section, and by 1888 it was estimated that 680 acres of land had been covered with waste to a depth of from 8 to 25 ft, and amounting in all to 9 million tons, with a sulphur content of about 15 per cent. (Fletcher, 1888, p.258; Hardie, 1950, p.106, gives for 1891 the figure of 10 million tons of waste on 500 acres in the Ditton and Widnes Marsh districts; see also Rhodes, 1909.)

CHAPTER 10

The Ammonia Soda Process

On 1 January 1865 production of soda ash by an ammonia soda process began at Solvay's Couillet works. The Winnington, Cheshire, works first made ash in the spring of 1874. In the suceeding 100 years Northwich was to remain unchallenged as the centre of the British ammonia soda industry, a position attributable not so much to unrivalled suitability as to the excellence of operations which were guarded by a sustained policy of eliminating rivals.

Solvay's process represented a considerable economy of material input, and therefore of material movement, as compared with Leblanc's—in 1872 in Mond's estimates approximately 130 cwt of material per ton of 56 per cent soda ash as against 170 cwt. Fuel consumption was about half as great as in the Leblanc process, but consumption of limestone and salt was heavier, and the latter was ideally in the form of brine. Early commercial estimates all confirmed that the Solvay process should give low production costs in comparison with those of Leblanc and it was an encouragement that by 1872 Solvay had declared a dividend of 100 per cent. A salt-field location was desirable, but low operating costs and high profit margins suggested that in other respects choice of location was not important. This is significant for it throws some light on the less than optimum locations that were chosen in some instances, and emphasizes by contrast the care, and also the good fortune, which attended the location of the first British works (Table 37).

Ten years after production began at Couillet, the works made one-quarter of the soda ash consumed in Belgium, but so little was known in Britain about these operations that the Solvay process was ignored, and when reported in the technical press the name of the firm was spelt Soldet (Vincent, 1874, pp.476–7). In 1873 at the Vienna International Exhibition, Solvay was awarded a gold medal, partly on the grounds that he had cut the loss of ammonia in the process to 2 per cent—though Weldon, who reported this, estimated that the true loss was probably nearer to 8 per cent (ibid.). Solvay's own choice of location was certainly defective, though limited by the mineral endowment of Belgium as well as by his own perception. In 1861 he had built a small factory in Brussels for the industrial application of his process. In 1864 he began the works at Couillet on the Sambre just east of Charleroi. There was coal and carboniferous limestone in the immediate vicinity, but, as Belgium lacks salt deposits, Solvay had to import rock-salt and dissolve it. In 1869 production was tripled, and four years later, recognizing the advantages of direct access to brine, Solvay built the much bigger works at Dombasle on the great salt-field south of Nancy and well placed to obtain coal by rail from the

Table 37. *Estimated production costs of Solvay and Leblanc processes, 1872*

	Mond (1872)		Mond (1972b)		Brunner (1872)	
	Leblanc at Widnes	*Solvay at Winsford*	*Leblanc at at Widnes*	*Solvay at Winsford*	*Leblanc at Widnes*	*Solvay at Winsford*
	£ *s. d.*	£ *s. d.*	£ *s. d.*	£ *s. d.*	£ *s. d.*	£ *s. d.*
Salt	12 3	1 6	12 3	1 0	12 3	1 6
Slack	1 8 6	13 8	1 15 2	13 8	1 15 2	17 3
Coke	—	5 0	—	5 9	—	5 0
Limestone	11 4	14 4	11 4	8 10	11 4	14 4
Pyrites	1 10 9	—	1 10 9	—	1 10 9	—
Nitre	8 0	—	6 6	—	7 0	—
Sulphate of ammonia	—	1 16 0	—	1 16 0	—	1 16 0
Total material costs	4 10 10	3 10 6	4 16 0	3 5 3	4 16 6	3 14 1
Wages	1 12 0	15 0	1 12 0	15 0	1 12 0	15 0
Maintenance	14 0	11 0	14 0	11 0	14 0	11 0
Packages	18 0	1 2 0	18 0	1 2 0	18 0	1 2 0
General expenses and depreciation	1 5 0	14 8	1 5 0	14 8	1 5 0	14 8
Total production costs	8 19 10	6 13 2	9 5 0	6 7 11	9 5 6	6 16 9
Freight to Liverpool	2 0	3 6	2 0	2 0	2 0	3 6
Royalty	—	8 0	—	8 0	—	8 0
Grand Total	9 1 0	7 4 8	9 7 0	6 17 11	9 7 6	7 8 3

Saar. By 1878/9 Couillet produced about 5,000 tons of soda ash annually and Dombasle 20,000 tons (Spons, 1879, p.316). In the next few years a number of works were built on the Continent and in the U.S.A. (Fig.16).

In the spring of 1874 an important meeting was held at the Society of Arts in London at which Vincent (1874, *passim*) gave a paper 'On some recent processes for the manufacture of soda'. One of his objects was to show that 'scientific correctness alone is insufficient to secure success'. He went on to say:

> The failure of only too many chemical processes arises from the want of technical skill on the part of their devisers, and also, frequently, from the abundance of technical skill bestowed on the operations they are desired to supersede. This is emphatically the case with those processes which have been designed with the view of replacing our present mode of soda making.

He noted that at the end of the eighteenth century the price of kelp containing an average of 3 per cent of alkali was £11 a ton, which would render one ton of soda ash worth £180, and therefore 'a very large premium for the introduction of a new process' was offered, but by contrast with that 'the present price of soda is £8 per ton, and economy in manufacture has been so studied in every branch of the process, that the premium for a new method is so small as to constitute one of the greatest difficulties that those reactions proposed to supplant our present mode of soda making have to contend with'. Vincent concluded, 'The alkali trade still awaits the coming man, and if one may venture an opinion, it is not a new process that is required so much as improvements upon the old.' In discussion Walter Weldon pointed out that the

Fig. 16. European ammonia soda plants 1883 *Source:* Weldon, 1883

ammonia soda process had been at work for the last seven years in Belgium, but he added that the loss of ammonia was still serious and that 'the process might have an immediate chance of being adopted if it were not for its extreme slowness'. This seemed in Weldon's opinion the most serious objection to it, though if the loss of ammonia was 8 per cent, this alone was 'very nearly if not the whole of the ammoniacal liquor produced by all the gas works in Great Britain'. Weldon concluded, 'For his own part, he was extremely sorry that this was the case, because the Ammonia process was a very charming one, being beautifully simple, yielding an exceedingly pure product of a perfectly white colour, in fact superior to anything which could be produced by the old process; above all it produced no nuisance.' Wincent's paper was published in April 1874; in the same month the first soda ash was produced at the Winnington works of Brunner and Mond. The 'coming man' had arrived.

The origins of Brunner, Mond & Co. and of the Winnington works

Hagerstrand (1973, p.83) has observed, 'Workable opportunities are really only those which are known, to the decision maker, which have acceptable internal characteristics and which permit entrance at the right moment.' These conditions were ideally met in the case of the establishment of the Solvay industry in Britain. John Brunner joined John Hutchinson's Widnes alkali works in 1861; shortly afterwards he met Ludwig Mond. It was Brunner's commercial ability and financial skill which supported the technical genius of his partner. Brunner was a man of 'optimism, courage and idealism'. By contrast Mond was 'ebullient . . . a veritable dynamo that only death could harness'. He was 'a hard task master, but none was driven so hard as himself', gifted, tireless, and never contemplating defeat. Once it was said in excuse of the failure of one man in a task that he had done his best. Mond observed, 'Heaven help the man who does his best and fails; there is no other hope and I have no use for him.' (McGowan, 1948, p.1; Watts, 1918, pp.320–1.) Born at Cassel in 1839, in the mid-1850s he studied chemistry at Marburg and then under Bunsen at Heidelberg. After working in a small Leblanc works at Ringenkuhl near Cassel, at which he was appalled by the sulphur lost in the operations, and having had other chemical works experience, he came to England in 1862 to work at Hutchinsons. Between 1864 and 1867 he built and managed Leblanc works at Utrecht, but in 1867 returned to Widnes. In 1861, with additions in 1862 and 1863, Mond had patented a process to recover sulphur from Leblanc tank waste, though this was eventually wholly superseded by Chance's process. Through his work on sulphur recovery Mond built up a close knowledge of the Lancashire Leblanc industry. In 1864 he had visited most of the Tyneside alkali works in an attempt to interest them in sulphur recovery, and at the end of the 1860s made trips to St. Rollox and to Netham works. Brunner, as head of Hutchinson's office staff, had access to detailed cost figures. In fact, at the end of the sixties, when Mond's energy was seeking new employment, the two men were in a highly favourable position to judge prospects in the chemical industry. By 1870 Mond was planning a sulphur recovery works to serve a group of alkali plants in Lancashire, no one of which was big enough to support such a works by itself, but this came to nothing. By summer 1871 he had decided to enter chemical manufacture in Britain and considered at least four different schemes to be entered in partnership with Brunner—a fertilizer plant near London, the manufacture of ammonium sulphate from gas-works liquor, a soda crystal factory, and one for production of alkali and bleaching powder. Early in 1872, having learnt about Solvay's process, he wrote to his parents in terms that showed the uncertainty of his plans but suggested the germination of the ideas that were to revolutionize the British alkali trade.

There is not much prospect for new business, as all the soda works are excessively busy, and earn so much money that they pay no attention to matters of minor importance in proportion to their own business [the sulphur recovery process is implied]. Consequently I am trying if possible to create an independent undertaking for the manufacture of sulphur, and if I can succeed with this I shall participate in it. But there are considerable difficulties to be overcome, especially the securing of the raw material for many years to come without paying for it and the acquisition of suitable land to deposit the lime residues; the negotiations require time and patience. I have in the meantime postponed my plan for the erection of a soda works [Leblanc] in order to wait for the results of several new processes, which, in all probability, will much simplify this manufacture, and especially the plant required but which have not yet been sufficiently exploited so that I could have dared up to now to adopt them. . . . It is not improbable that within the next few days I shall make a short journey to Belgium. (Watts, 1923, pp.5–6.)

In April, 'after a very close examination of the works and the books of the company', Mond came to an agreement with Solvay. The Solvay process licence was dated 23 September 1872 and under it the licensees agreed to build an English Solvay works within a year and to have it in production within two. Solvay agreed not to make arrangements with other British firms within this period, and, if such agreement were subsequently made, to charge a royalty of 20*s*. a ton as compared with 8*s*. for Mond's plant. By December 1872 Mond was jubilantly confident. He believed that even when existing Leblanc works were unprofitable, the Solvay process would give a net profit of 50 per cent, and with current inflated prices it would probably be over 100 per cent. He wrote exultantly, 'It is really the most lucrative and extensible invention which has been discovered by chemical manufacturers for the last eighty years, and I consider it to have been a great fortune for it to have come into my hands.'

It must have been clear to Mond that his proposed works would have important advantages not only over Leblanc works but also over Couillet. He planned for a large production. In 1872 Couillet was making about 10 tons of soda ash daily from two strands, (though by 1873, according to Kuhlmann at the Vienna Exhibition, it was making 15 tons). Mond's designs were for 20 tons from one strand. (*J.S.C.I.*, 29 Sept. 1885, p.527.) By late December 1872 Brunner and Mond had made the vital first step towards the high capacity which Mond planned, they had found a wealthy individual to act as a sleeping partner—in fact a lender rather than a partner. Charles M. Holland, the successful Manchester civil engineer, provided an initial capital of £5,000. Mond wrote home:

> Having concluded this, I have got over the greatest difficulty about the new undertaking, and now I am very busy in order to get over the second great difficulty, the acquisition of the necessary site. It seems that I shall have more difficulties to find it than I anticipated. No-one wishes to give a site for soda works, as they do not believe that the new method will not have as many drawbacks as those experienced with the old method, but also in this respect I have made progress. (Watts, 1923, p.7.)

Locational choice

In choosing a location for their plant Brunner and Mond acted rationally but did not look far afield. A key requirement of course was brine, but there is no evidence that they ever considered the Tees salt-field district where the brine had been proved ten years earlier—it may be that this was due to knowledge of the great depth of the salt, and also perhaps to the fact that they were oriented towards the Mersey Basin. Locational choice within the Mersey Basin had recently been affected by improvements to the railway. The completion of the Runcorn Gap Bridge by 1869, for instance, shortened the rail haul on St. Helens or Wigan coal to Runcorn and to the Cheshire salt-field, and shipments of finished products from there to Liverpool were also eased as a result of the cutting-out of the roundabout route through Warrington.

Brunner later recalled the first stage of their search: 'When we had decided to adopt the Solvay process, Mond and I went from Widnes to Hartford station [2 miles south-west of Northwich] on May 4th 1872, and thence walked up the Weaver valley as far as Winsford, to survey the ground with the idea of choosing a site on which to establish our works.' Mond wanted to accept a leasehold site at Winsford from Falk, a salt manufacturer, but Brunner refused to build on leasehold land. Nevertheless, in their early estimates of production costs, both Brunner and Mond compared Widnes Leblanc costs with Solvay production costs at Winsford. Winsford was a good location for road and for rail transport: the Shropshire Union Canal, with its connections to the Trent and the Mersey, was only three or four miles away by road. The Weaver itself had been widened and deepened in 1867/8 and was navigable up to Winsford for vessels of 120 tons; 200-ton vessels could reach Northwich. (Imperial 1872.)

On a second visit, instead of following the Weaver valley, Brunner and Mond struck across the fields towards the little community at Anderton where the Trent and Mersey Canal linked with the Weaver. On the walk they were encouraged by the group of salt-works near Winnington and impressed by Winnington Park. 'It was freehold, it had a long frontage on the navigable river Weaver, and, therefore easy access by water to the Liverpool docks, and it had railway connection with the Cheshire Lines at a convenient level. Mond said no doubt we could get brine there,' was how Brunner recalled the situation (Watts, 1923, p.22). Tangye (1948, p.1) gave a variant of this account from the statements made in his hearing by Mond at the works anniversary celebrations of 1898, 'As we approached the canal lift [the hydraulic] he told me how he and Mr. J. T. Brunner took a dog cart from Farnworth one Sunday and came over to look at this celebrated engineering feat: and leaned over the railings of it looking down over the Winnington Hall estate and the brine land towards Northwich.'

On 6 October 1870, Winnington Hall with 675 acres of land had been offered for sale but there had been no takers (I.C.I., 1951, p.33). Two years later Brunner and Mond decided that this should be the location for their plant. Though they wanted only a small part of the estate and feared that the purchase price was in fact beyond their means, early in 1873 they succeeded in purchasing it from Lord Stanley for £16,000. On 28 February Brunner Mond & Co. was formed.

Mond became increasingly pleased with their choice of site and location: 'It is not only a fine property, but it is also in an incomparable position for our works, which offers us most substantial advantages towards any possible future competitors' (letter written to Germany in April 1873 from Widnes: Watts, 1923, p.22). In fact it was not in all respects what they had expected (Wallwork, 1967, pp.172–7). Success was due much more to business organization and efficient production techniques than to choice of location. What seemed the ideal sources and methods of delivery of raw materials in the winter of 1872/3 had to be changed in the light of experience. Brunner and Mond had intended to use the River Weaver for assembly of materials and for dispatch of products, but material availability and price forced them to turn to sources of supply that involved rail transport via the branch line (opened in 1866 to the Winnington–Anderton group of salt-works). St. Helens area coal proved 1*s*. a ton dearer than that from the Potteries and North Wales, and so the latter sources were preferred. By contrast, quotations for North Wales limestone were unacceptably high; consequent inquiries sent to the Buxton area in September 1873 resulted in orders being placed in October for 600 tons of limestone monthly at 4*s*. 5*d*. a ton of 21 cwt with the Tideswell and Millers Dale Coal Company. Arrangements were made to buy some ammoniacal liquor from firms for which water shipment was practicable—for instance, Thomas Hadfields of Liverpool—but there were also early contracts for liquor supplies from as far away as Sadlers of Middlesbrough. Salt supplies were sometimes precarious in the early days, but in 1878 two bores were put down in the south-eastern part of the estate, the only part containing salt and lying east of the Winnington fault, whose existence had not even been suspected when the Winnington site was chosen. (Wallwork, 1967, p.175; I.C.I., 1948, p.6.)

Processing problems proved greater than problems of supply, and here the genius of Mond proved far more significant than the choice of location. The steam engines were faulty and much other apparatus, though based on Solvay's plans, broke down; it was saved and improved by Mond, who for the first eighteen months spent most of his time in the works, even going to the length of having his bed on a platform over the engines. Despite his efforts, long after the first soda ash was produced in April 1874 it proved difficult to attain the high levels of production that had been planned (Table 38).

Table 38. *Brunner Mond & Co. production, 1874 and 1875*

	Output (tons) final six months of year	*Value of output (£)*	*Cost per ton (£. s. d.)*
1874	575	9,518	16 0 0
1875	1354	14,440	10 6 0

Note: The 1874 selling price was £10 to £11.

Source: Brunner Mond & Co., Winnington Works Balance Sheets.

On 14 August 1875, after almost eighteen months of running-in difficulties, Mond wrote to Charles Holland:

> We are getting on pretty well and regularly at the works, keeping up our make to rather better than 50 tons a week on the average, but my expectation to raise this to 60 tons has not yet been realized on account of a number of small difficulties with the plant here and there which have given me a great deal of work and trouble. I trust, however, that the next four weeks will open an era of better and more comfortable working as these small impediments have now been almost all removed. (Mond, 1875.)

This was far too optimistic; two weeks later he wrote urgently to Belgium: 'My dear Solvay, I have just this moment telegraphed you "Events entirely unforeseen prevent our honouring your bill. Send 3,000 francs or allow us to dispose on you".' They were asking 'once more' for financial help, and Mond added, 'We have not had a very good make this month. We only made about 195 tons.' (Mond, 1875.) At this stage interested outsiders were ready to write off the ammonia soda challenge as it had proved commercially unsuccessful for the previous twenty-five years. During 1875 Weldon reaffirmed his opinion that 'the process is so slow, and the apparatus so expensive, that it could not economically be used' (Cohen, 1956, pp.274–5). In fact, things were already swinging in Mond's favour. In almost nine months of operation in 1874 they lost over £4,000; in 1875, in spite of their difficulties, they broke even, and in 1876 were able to pay off most of their debts. In 1877 their plant was doubled and in 1881 the firm undertook further major extensions (Table 39).

Table 39. *Weekly output of alkali at Winnington* (tons)

1874	23.2
1875 (first 6 months)	43.5
1875 (final 6 months)	52.1

Sources: Brunner Mond & Co., Winnington Works Balance Sheets; Watts, 1923, p.31.

By 1880 Weldon had so far amended his previous judgement that he observed, 'The consideration that constitutes the chief ground of hope of the makers of Leblanc soda for salvation against the Ammonia Soda process, is that the world requires chlorine as well as soda' (Quoted Cohen, 1956, pp.274–5). In 1881 Brunner and Mond were able to sell their plant to a limited company for a nominal capital of £600,000. By the beginning of 1883, with Leblanc works making no profits on soda ash, Northwich was earning its operators £1 a ton profit. In 1880 the Solvay process turned out only one ton in every thirteen of British soda ash; six years later it turned out every third ton. In 1888 British Leblanc works used 585,000 tons of salt, ammonia soda works 212,000 tons; by 1894 the respective figures were 434,000 and 362,000 tons, and in 1896 361,000 and 431,000. The margins of profitability of the Solvay process had narrowed since Brunner and Mond had made their cost comparisons in 1872, but their costs remained lower (Table 40).

Table 40. *Production costs of Leblanc and ammonia soda process, 1894* (per ton)

	Ammonia Soda 58% ash £ s. d.	*Leblanc Soda 55% ash* £ s. d.
Saltcake	—	16 8
Brine and pumping charges	9	—
Limestone	4 6	6 6
Slack	9 0	13 9
Coke	1 3	—
Loss of ammonia	6 0	—
Wages	14 3	14 0
Management, laboratory, and office expenses	3 6	2 0
Repairs, wages, and materials	5 6	4 6
General charges	14 0	4 6
Total	2 18 9	3 1 11

Source: *C.T.J.*, 3 Nov. 1894, p.269.

Rival ammonia soda firms

No other attempt to produce ammonia soda in Britain was a long-standing or marked success. In 1891, discussing the industry and plans for new works, the leading trade journal noted 'it is true that other works have plant for this manufacture, but these are toys in comparison with the colossal undertaking at Northwich'. Firms that entered this trade would need skilled technologists and managers to give them a chance to compete. (*C.T.J.*, 3 Jan. 1891, p.2.) None had the unique combination of Mond's technical flair and Brunner's commercial acumen. Brunner and Mond entered into agreements for price arrangements in order to avoid or limit competition. In almost every instance this led eventually to their buying and scrapping or buying and re-equipping the works of their rivals (Fig.17).

Three ammonia soda works were built outside Cheshire. The first was put up at Port Clarence, Teeside, in 1884. By 1900 it had been demolished. Neil Mathieson built an ammonia soda unit at Marsh works, Widnes, in the late 1880s. It had an annual capacity of about 15,000 tons in the early 1890s but

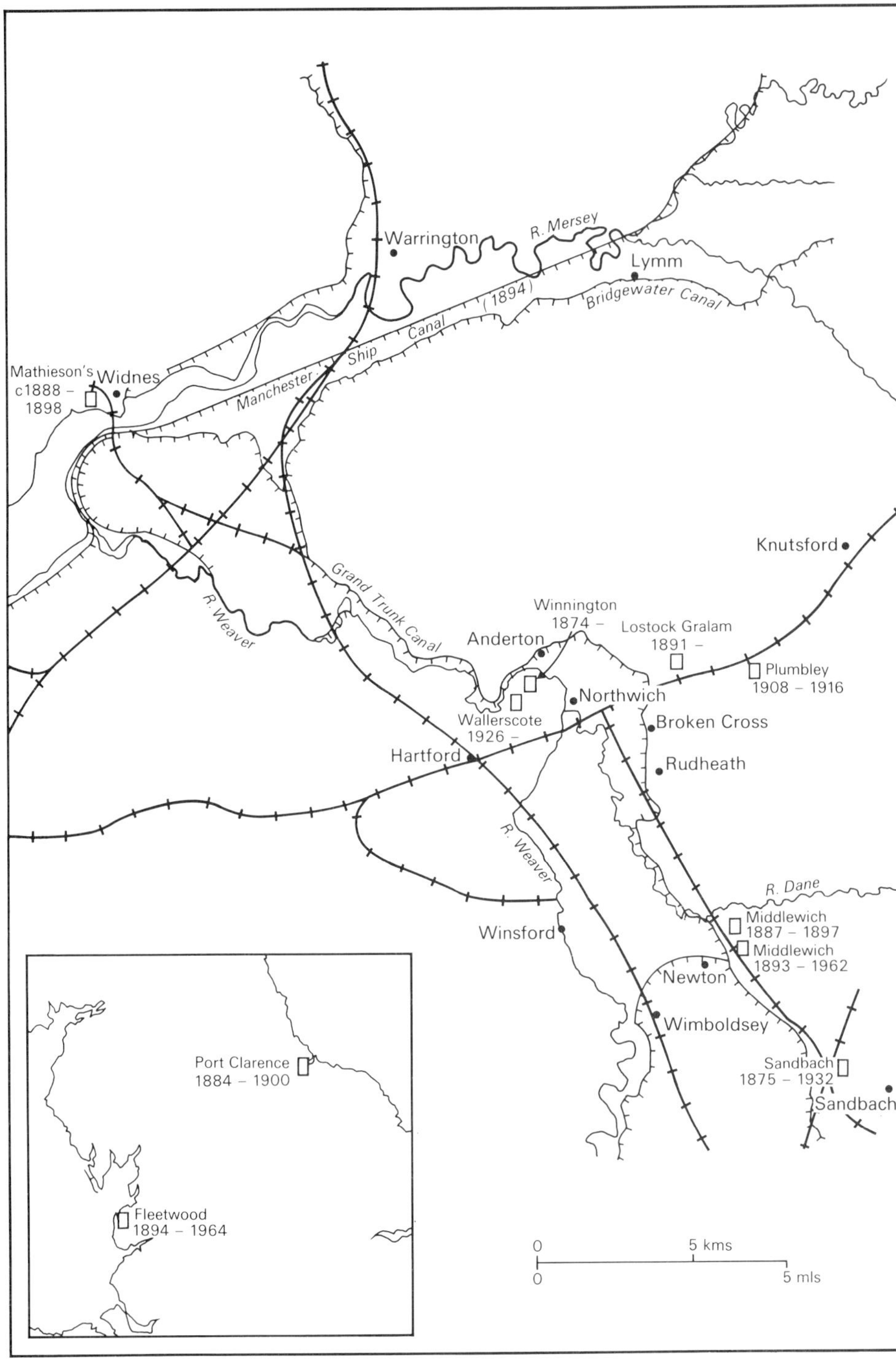

Fig. 17. British ammonia soda works *Sources:* various

was closed in 1898. By 1891 the United Alkali Company had decided to build their own ammonia soda works at Fleetwood. It survived until 1964.

In addition to Winnington, six ammonia soda works were built in the Cheshire salt-field; two survived a century later. In almost every instance failure of the others was due to incompetence in management or technical deficiency rather than a non-viable location. Sandbach works was built by a group led by James Richards, a man whom Brunner had tried to bring into the original partnership. From the start Sandbach was burdened by the 20*s*. a ton royalty to Solvay. It provided troublesome competition but was not profitable to its own promoters. Many years later it was reckoned that the plant had been commercially viable when working but that it did not in fact operate more than two-thirds of the time, largely because of its very complicated construction, particularly in the ammonia stills (Mason, 1908, p.420). In 1877 it was arranged that Mond should manage the Sandbach works from the beginning of 1878, being paid £1,000 a year for management and 20*s*. a ton of output, the 12*s*. difference between the Sandbach and the Brunner Mond royalties being retained as profit. Output now moved rapidly upwards (Table 41).

Table 41. *Soda ash production at Winnington and Sandbach* (thousand tons 58% ash)

	Winnington	*Sandbach*
1874	0.8	—
1875	2.5	—
1876	3.3	0.9
1877	5.9	1.6
1878	7.7	3.6
1879	10.1	5.6
1880	12.6	6.2
Total	42.9	17.9

Source: *Proceedings of the Society of Chemical Industry*, 1881/2.

In 1881 the option to take over the Sandbach works was exercised, and by 1883/5 capacity was being doubled. Further additions were made in 1892. In 1882 Dr. Pick, who had managed Sandbach for Brunner Mond & Co., left the firm. Five years later he became associated with a rival concern, the Cheshire Alkali Company with capital almost wholly derived from residents of Crewe. This company decided to build at Middlewich, a town whose development had fallen behind that of its neighbours. Improvements which Dr. Pick was said to have made in the process were incorporated. (*C.T.J.*, 30 July 1887, p.77.) The plan was to make 12,000 tons of soda ash, 6,000 tons of caustic soda, and 5,000 tons of soda crystals a year, using the firm's own sources of brine acquired in the Newton salt-works in the suburbs of Middlewich. In terms of location, Middlewich was as good as Winnington. Early publicity spoke of a possible return on capital of 20 per cent, but six years after its formation the plant was in trouble with creditors, and within ten years, by arrangement with Brunners, was dismantled. Middlewich had used a combination of the ammonia soda and the Leblanc processes (sodium sulphate being made by the Leblanc process and converted into soda ash in the presence of ammonia). The scale of operation was large—some estimates put the capacity at almost 500 to as much as 600 tons weekly (or more than 50 per cent over the level of Winnington and Sandbach combined as late as 1880). The Middlewich operating rate was, however, low, reaching not quite two-thirds of this potential. (*C.T.J.*, 2 June 1888, pp.346–7; Mason, 1908, p.421.) This poor performance resulted from defects in chemical engineering and a lack of the selfless attention to detail that Mond had lavished on Winnington in the earliest years. Brunner Mond were declaring dividends of 50–60 per cent or better while others went down in the same market conditions. In 1893, when the creditors of the Cheshire Alkali Company met in Crewe, it was agreed to defer payment (liabilities were heavy) while new machinery was procured to lower production costs. (*C.T.J.*, 24 June 1893, p.402.) As the *Chemical Trade Journal* (17 June 1893) warned yet again, the ammonia soda process was simple in principle but complex in detail and new companies did not take this sufficiently into account. Either the advice was not taken or perhaps Brunner Mond were now so firmly established that no new firm could make its way and flourish. Four years later Middlewich was closed and 'entirely scrapped, an utter failure' as one of Mond's young men recalled (Tangye, 1948). In 1890 the Middlewich Salt and Alkali Company was formed to work salt and make alkali at Newton. It did not get beyond the acquisition of land and a few experimental bores; within two years it was bankrupt. (*C.T.J.*, 26 Mar. 1892, p.212.)

In 1891 two displaced managers from the United Alkali Company, backed financially by interests from the cotton industry, built at Lostock Gralam (Irvine, 1958, p.22). The firm, Bowman Thompson, operated ammonia soda works, but it also operated a partial Leblanc works to produce sulphuric acid, saltcake, and bleaching powder. It sold hydrochloric and sulphuric acid and saltcake as well as alkali widely throughout Britain (Fig.18). It was hit badly by the depression of 1893, and, in trying to counteract this by reducing wages, ran into labour troubles. In 1895 it entered into agreements with the United Alkali Company and with Brunner Mond & Co. to fix prices. The aim, as Brunner put it, was to end the 'thoroughly unwholesome state of affairs' in which alkali prices had fallen below costs, but not to raise prices to a level that would encourage new entrants to the trade. 'Looking ahead, as the directors of the company always felt themselves bound to do, they were very decidedly of the opinion that they ought not to put themselves in such a position as would encourage fresh competition in the future. Therefore he thought all parties were agreed that they ought not to raise the market price so as to encourage undue competition.' (Brunner, 1896, p.145.) In 1898–9 Brunner Mond & Co. paid 30 per cent dividends, whereas Bowman Thompson never paid more than 10 per cent. F. H. Bowman told his annual meeting in the

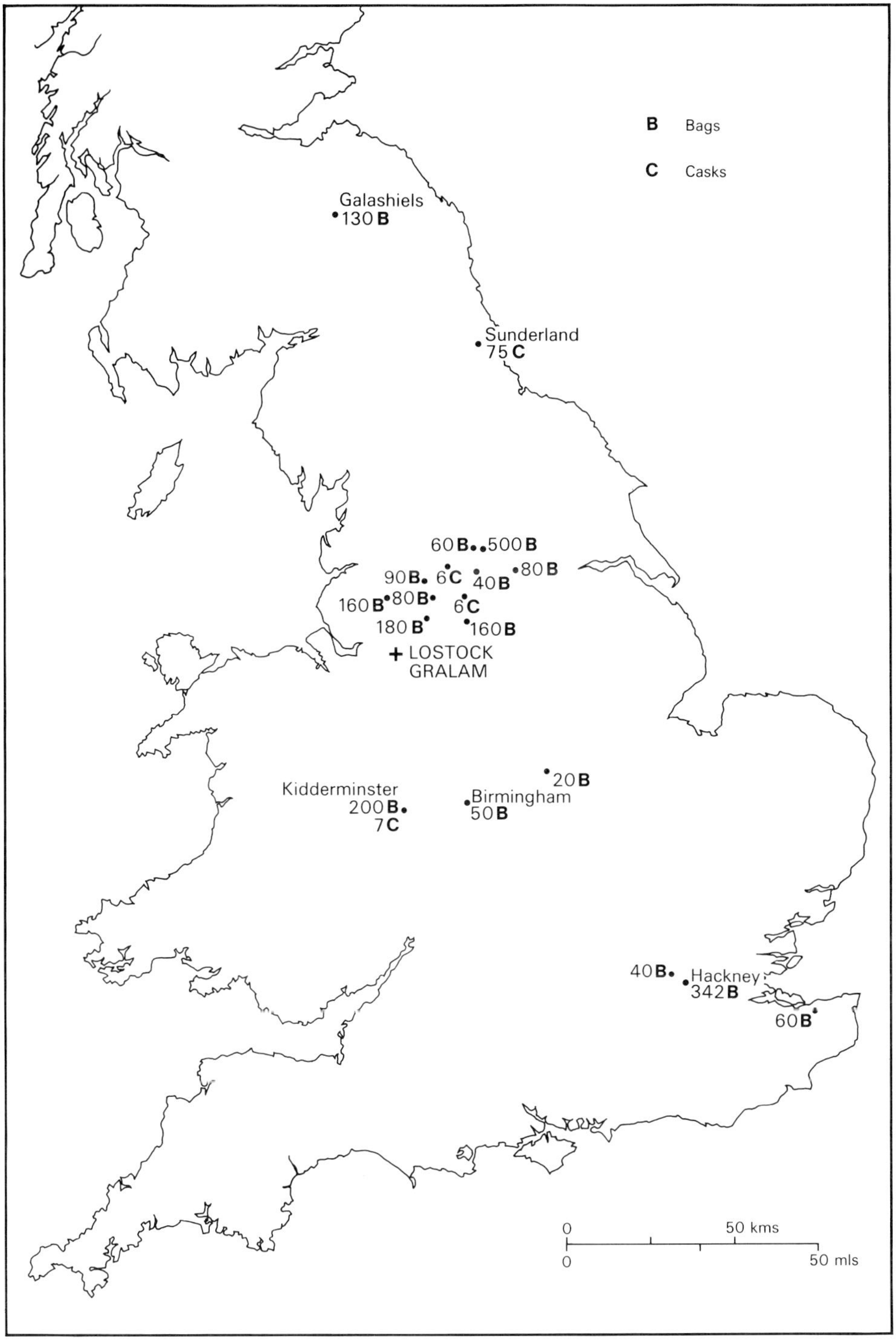

Fig. 18. Deliveries of alkali from Lostock Gralam 1896 *Source:* Brunner Mond, Correspondence

spring of 1900 that the works was in a state of high efficiency in all departments, but he also observed that if the tripartite agreement was not renewed, £100,000 more capital would be needed, 'in order to compete in the open market with any chance of success'. At that stage Lostock Gralam employed 300 and produced 1,500 tons of finished chemicals weekly. The plant was sold to Brunner Mond that year for £350,000. (*C.T.J.*, 19 May 1900, p.328; 9 June 1900, p.378.) The sale agreement involved a commitment that the Bowman directors would not be concerned in the production of ammonia soda or its products within fifteen years in the United Kingdom or in any British possession. Mond visited their newly acquired works and was reported to have been delighted with their 'fine' vitriol plant and 'excellent' Weldon chlorine unit, but 'not a bit interested but amused at their Ammonia Soda attempt'. Lostock Gralam was 'an unscientific failure over Ammonia Soda and was taken over . . . largely on account of its vitriol and hydrochloric acid facilities and the brine supply' (Tangye, 1948). Norman (1931) suggested of Bowman Thompson that 'but for wine and women they should have made a fortune'. Mond reported to his next Annual General Meeting:

> with reference to the works which they had acquired, they found a portion of these works for the production of vitriol, saltcake and bleaching powder to be well laid out, and, while they found them in most deplorably dilapidated condition, they had been able in a short time to be able to put them in thorough repair, and they were now doing as well and manufacturing as cheaply as any similar works in the country. As regards the Ammonia Soda works which they acquired, they found them to be so badly laid out and so deficient in detail, that they determined to stop them for a time and had not yet decided whether it would be worthwhile to rebuild or reconstruct those works, or to take them down altogether. (*C.T.J.*, 8 June 1901, p.526.)

Meanwhile Lostock Gralam was entered in the balance sheet at breakdown value (ibid.). It was reconstructed and ammonia soda manufacture was resumed in 1907.

In 1893 another works was built at Middlewich, by G. L. Murgatroyd to the plans of Daguin of Varangeville. The greater success of this operation was attributed by some to 'careful management' but was also due in part to Murgatroyd's engineering, for the plant had some splendid apparatus. (Watts, 1923, p.41; Dick, 1973, p.21.) Opportunity to absorb it came on Murgatroyd's death in December 1894 at the age of 44. Within eight months Brunner Mond & Co. had taken over. They closed the works, but, after reconstruction, reopened them in 1897, and operated them for another sixty-five years. Long afterwards, Alberty Tangye, who had joined Mond as a young man in 1885, recalled that, given the Murgatroyd equipment, 'it was well to absorb it' (Tangye, 1948). The same attitude seems to have characterized the approach to the final ammonia soda works built by rivals in Cheshire.

In 1907 a very thick deposit of salt was discovered at Plumbley, and in 1908 the Ammonia Soda Company established works just across the railway line from Brunner Mond & Co.'s works at Lostock Gralam. The plant was 'of poor and antiquated design' and therefore was reckoned incapable of making a profit in the competitive conditions then existing. (Watts, 1923, p.44; *Brunner Mond Record*, July 1926, p.5.) This seems a strange assessment given the composition of the board of the new concern. The leading shareholder was Ivan Levinstein of the great Manchester dyestuffs firm, and the biggest group of shareholders was made up of Chamberlain and Nettlefold family interests. There were site difficulties: in 1923 Watts recalled, 'it had no adequate water supply and no facilities for the disposal of effluent liquors'. Fifty years later Dick (1973, p.34) showed that the hostility of Brunner Mond & Co. played an active role: 'The site was suitably placed for raw materials but lacked adequate long-term provision for waste disposal since Brunner Mond had purchased the surrounding land.' Moreover, he continued: 'Brunner Mond were constantly vigilant for any sign of river contamination which would threaten their own water supplies and finally the Ammonia Soda Company went into liquidation in 1916.' That year Brunner Mond purchased all but a few of the ordinary shares. In March 1918 Plumbley was stopped. In August 1919, at an extraordinary general meeting, it was resolved that the company be wound up, a question 'whether Brunners did not stand to benefit substantially by the liquidation of the Ammonia Company' being ruled out of order. (General Chemicals, 1963.)

There seems to have been a danger that in all these cases, in absorbing the opposition, Brunners would increase not only their capital commitment but also operating costs. They met this difficulty by writing off and demolishing some plant, but where they retained operations they recognized the need to cut costs. One way of doing this was to reduce wages at outlying works. In the late autumn of 1897 they called the men at Sandbach and at Middlewich together and declared that owing to depression they could not continue those works unless the men submitted to wage cuts of 7–15 per cent. If the men accepted, the works would run full-time. (*C.T.J.*, 11 Dec. 1897, p.375.) Some months later Richards of Sandbach works wrote to Northwich. He reported that his foremen had requested an increase in wages but that he had reminded them that the conditions on which it had been decided the previous year to carry on the works were that wages there should be lower than those at Winnington. (Brunner Mond & Co. records, I.C.I., London, 12.1.6a.) Although their other works survived, Winnington remained dominant (Table 42.).

Table 42. *Brunner Mond soda ash capacity, 1913* (tons per day)

Winnington	600
Lostock Gralam	200
Sandbach	150
Middlewich	150

Source: Brunner Mond & Co., Estimated Fuel Requirements, 1915, in I.C.I., Runcorn Records.

Apart from Neil Mathieson at Widnes and the United Alkali Company at Fleetwood, Leblanc firms were remarkably unwilling to enter the ammonia soda trade. At some stage it seems that the Runcorn Alkali Company planned an entry—a site was chosen at Lostock Gralam—but the commercial manager, Wigg, would not launch into development away from the Runcorn area (Norman, 1931). In 1901 Gossages and others from Widnes were said to be contemplating works on the new brine-fields at Rudheath/Broken Cross near Northwich (*C.T.J.*, 14 Dec. 1901, p.543; 28 Dec. 1901, p.578). In 1875 there was a proposal within the firm of Chance & Hunt of Oldbury to replace the Leblanc process with an ammonia soda plant. Objections were raised about the price of salt at Oldbury and the dearth of ammonia (salt costing 6*d*. in Cheshire cost 13*s*. 3*d*. a ton in Oldbury (Chance & Hunt History, n.d.)). After taking an interest in Chance and Hunt, Brunners reported of this ammonia soda scheme that 'after full consideration it was deemed inadvisable to make the venture at all events at Oldbury' (*Brunner Mond Record*, 1926, p.11). This summary of the Chance and Hunt decision not to go into the business was perhaps too concise: Tangye claimed to have heard Alexander Chance remark in 1885 that if he had been twenty years younger he would have started up an ammonia soda works at Droitwich. In 1890 Netham obtained a licence to operate the Hawliczek Ammonia Soda process, but in the autumn United Alkali Company was formed and the project was not carried further (I.C.I., Netham).

Brunner Mond & Co.: later developments

In 1881—by which time, as Mond put it, 'We are no longer making chemicals, we are making money'—Brunner Mond & Co. became a limited company. Thereafter major extensions were set in hand at Winnington and Sandbach (Fig.19), and the pre-eminence of the new concern was consolidated and extended. It enjoyed major technical advantages thanks to its close association with Solvay, with whom friendly relations were maintained. (Watts, 1923, p.51.) The scale economies were also important. When Winnington came into production in 1874, each element of the plant made 20 tons of ammonia soda daily. Each of the new elements brought into production in 1885 was of 60 tons daily capacity, which effected 'a great saving in labour, repairs and first costs, as compared with the older plant' (*J.S.C.I.*, 29 Sept. 1885, p.527). Integration moreover led to further economies.

In 1879 Mond resolved to study how to make ammonia more cheaply. It was ironical that the early patents and operations had been undertaken at a time when 'the ammoniacal liquor of the gas works could have been had for the carting away', yet the process was not worked until ammonium sulphate was £17 a ton (*C.T.J.*, 9 July 1887, pp.50–1). Ammonia supply was eased as the Scottish iron companies fitted by-product recovery plant to their blast furnaces—Bairds, the largest firm, could recover by-products from 300,000 tons

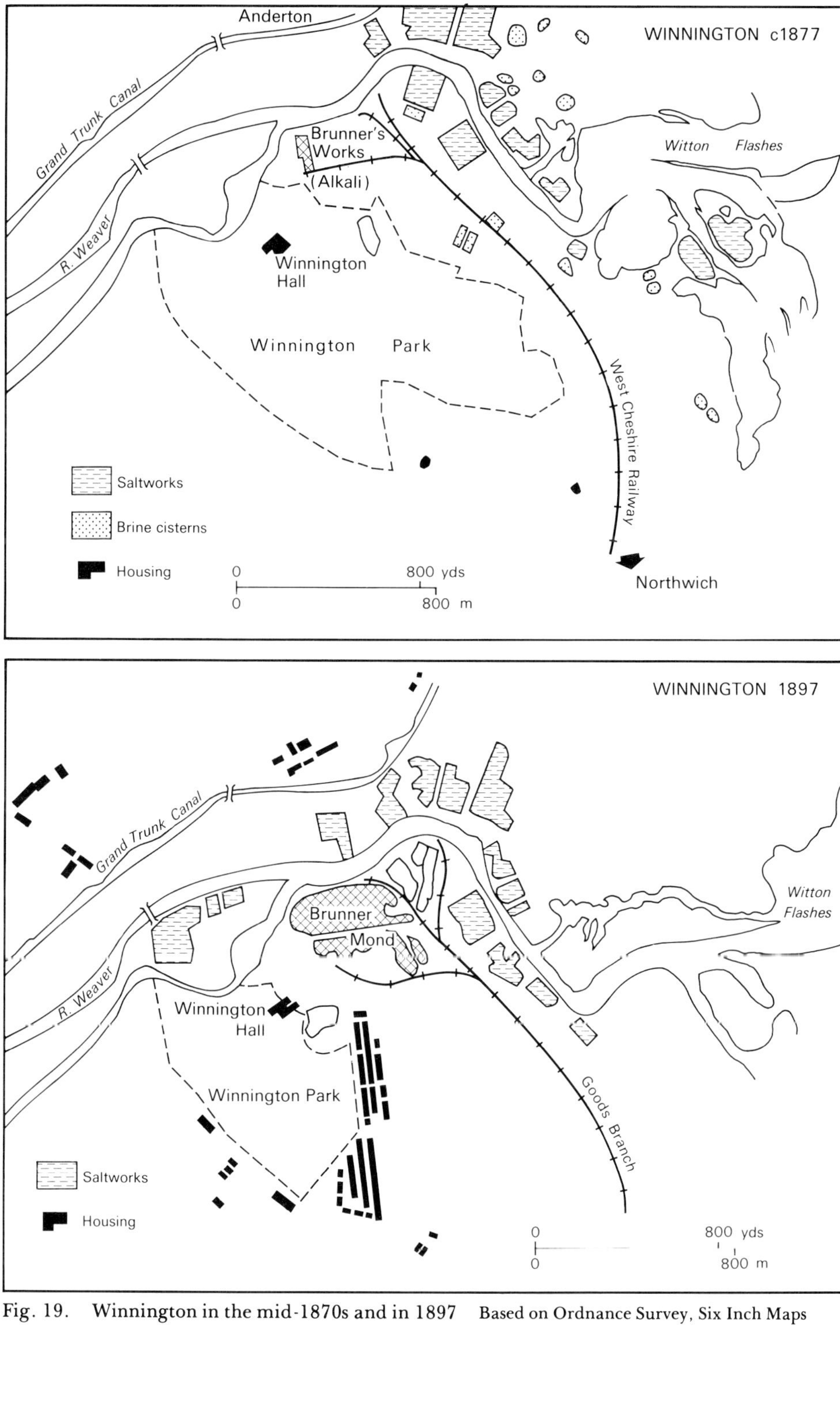

Fig. 19. Winnington in the mid-1870s and in 1897 Based on Ordnance Survey, Six Inch Maps

of coal by 1883—so that by 1892 fifty-seven out of seventy-seven blast furnaces at work in Scotland were equipped to recover by-products, including ammonia (*J.I.S.I.*, 1883, 2, p.698; *B.P.P.*, 1892, XX, p.13). In addition, Mond installed gas producers, and then, following up Solvay's work in by-product coking, installed the first Semet-Solvay coke ovens at Winnington in 1886. Not only did this assure Brunner Mond & Co. of ammonia supplies, it also provided them with additional profitable sales such as those of ammonium sulphate. For every ton of slack burned, Brunner Mond in the late eighties produced 66 lb of sulphate of ammonia. (*J.S.C.I.*, 30 Nov. 1888.) Other supplies were also gradually brought under their control. In 1873/4 their first limestone had come from Tideswell, Derbyshire, and by 1880 their demand was a major factor for the lime firms competing for their business. In November 1891 Buxton Lime Firms Limited was formed to merge all the old rivals, sixteen in all. In 1918 Brunner Mond & Co. absorbed Buxton Lime.

At the finishing end also, Brunners widened their interests. In 1894, in order to compete more effectively in the London market with U.A.C. and with increasing imports from the continent, they started a soda crystal plant at Silvertown. The following year caustic soda manufacture was begun at both Winnington and Silvertown.

A plan for Brunners to make ammonia soda in a new works on Teesside seemed likely to be realized in the mid-1890s, but was frustrated by agreement with U.A.C. Brunners were to erect one more ammonia soda plant in Britain. In 1912 land was acquired immediately west of Winnington, but wartime restrictions delayed building. Wallerscote works was operated from the end of 1926. Fifty years later, Winnington, Wallerscote, and Lostock Gralam survived as the only British ammonia soda units.

Brunner Mond & Co. had some difficulties with major buyers. In 1881 John Crosfield had become the first chairman of Brunner Mond & Company Limited, but in 1907, when Crosfields were given favourable contracts for the purchase of soda ash, they had to bind themselves not to make caustic soda or soda for general sale. Long before this, Lever Brothers had become the biggest soap producers in Britain—from making only 20 tons a week in Warrington in 1886, at Port Sunlight their output had risen by 1899 to 2,400 tons weekly. By then they were causticizing soda ash to make their own caustic soda. In 1907 a long-term agreement was negotiated for sales to Lever Brothers, but in 1911 the latter seemed to present a threat by their acquisition of salt lands at Lymm in the most northerly section of the Cheshire salt-field. Brunner Mond's reaction was to buy the next biggest soap firms, Crosfields and William Gossage and Sons of Widnes. Levers did not go ahead at Lymm, and in 1913 Brunners negotiated another long-term sales agreement, and in 1920 sold them Crosfields and Gossage. (Musson, 1965, p.240; Watts, 1923, p.43; Dick, 1972, pp.39–40.)

CHAPTER 11

Tyneside from Pre-eminence to Decline

By 1866 between Blaydon and the sea, Tyneside had a concentration of alkali works more or less equal to that from St. Helens to Widnes and Runcorn and containing some large, efficient, and successful works as well as many smaller operations whose weaknesses were soon to become disastrously obvious. Expansion continued until after the Franco-Prussian war, though at a slower rate than in the North-West, the Lancashire branch of the industry increasing its decomposition of salt by 38.1 per cent in 1866–74, the Tyne branch managing only 22.3 per cent. (Kingzett, 1877, p.75.)

Market changes

It was normal in Britain during the Industrial Revolution for industries originally located so that they could serve local needs to outgrow them. On Tyneside, as the alkali trade grew, local consuming industries stagnated and then rapidly declined. Exports became very important, though in the mid-1880s the Newcastle chamber of commerce estimated that 47 per cent of Tyneside's bleaching powder and soda products found a British market. (Allhusen, 1886, pp.264–5.)

In 1850 one hundred works in the North-East made glass bottles. By 1863 there were only forty-seven bottle works left and all six of the large crown glass works which had been operating on the Tyne twenty-five years earlier were derelict, although on the Wear Hartleys were making as much glass (but now sheet glass) as the six Tyne works together had made. In 1866, when the Lancashire alkali firms sold 32,000 tons of saltcake to the glass industry, Tyne firms' sales were only 2,400 tons. By the mid-eighties the number of bottle houses on the north-east coast estuaries had fallen to ten. Plate-glass making still flourished in South Shields and in Sunderland in the late eighties and speciality glass was being made in Gateshead, at The Teams and at Blaydon, but all over Tyneside smokeless cones bore witness to the collapse of a former staple of the river and a significant local outlet for alkali. (British Association, 1863, p.55; Armstrong, 1864, p.199; Mond, 1869, p.234; Clapham, 1878, p.161; Depression of Trade 1886; Stuart, 1889, p.235.) The men in the glassworks had a very compact union and used long strikes to enforce their demand for wage increases. The owners proved unable to resist, for the trade required special training and substitute labour was consequently hard to find. (*M.J.*, 20 Feb. 1886, p.221.) Belgian competition was also a heavy burden. By World War I there were some twenty-six glass-bottle-makers in the United Kingdom, none of which was in the North-East (Davis, 1913, p.349).

By the third quarter of the nineteenth century Tyneside had also ceased to be a major centre of soap-making. In 1862 only 6,000 tons were produced, roughly representing a market for no more than about one-tenth of that tonnage of soda, usually as caustic soda. Production of caustic on the river was just at that level. By the late 1870s one or two soap-works had been stopped and production was not much more than half the 1862 level. Ten years later production from two soap-works was 3,000–4,000 tons, a negligible proportion of a national output reckoned at 250,000 tons. (Clapham, 1878, p.161; *Proceedings, Society of Chemical Industry*, 1882, p.5.) On the eve of World War I, Hedley and Company of Newcastle and Davis of Birtley were the only two soap-makers in the North-East out of 184 soap firms in Britain (Davis, 1913, pp.405–7).

In 1861 the paper-maker Thomas Routledge of Sunderland introduced esparto grass as a substitute for rags. Alkali was needed to remove its silica and resin. (*C.T.J.*, 20 Jan. 1906, p.71.) In 1862 the United Kingdom imported 18,000 tons of esparto grass, of which half came into the Tyne, and by the late sixties the trade was such a growth industry that local shipowners were buying up once famous Australian and American clipper ships, and converting them for the coke/esparto grass trade. (Bell, 1869, pp.98–9; *C.G.*, 16 Jan. 1864, p.41; *The Times*, 4 Nov. 1871, p.6.) Gradually, however, the increase in the traffic justified the installation of baling equipment at the chief shipping points, carriage became less dependent on heavy minerals ballast, and the unloading of esparto was decentralized from the Tyne. Imports of the grass into the district rose from 18.4 thousand tons in 1863 to 35.9 thousand in 1873, but dropped to 23.1 thousand tons ten years later. Nevertheless, as late as 1882 Manchester paper interests complained that as most esparto came into either Newcastle or Cardiff they had to pay 20*s*. a ton on railway carriage. (British Association, 1889, pp.168–9; Sunderland, 1886, p.399; *Ship Canal Gazette*, 22 Nov. 1882, p.37.) For meeting the needs of British markets outside the region, the Tyne was at first advantageously placed, but as public inquiry exposed the discriminatory nature of earlier rail pricing, some of its advantages in this respect were removed. (Railways, 1881, Appendix, p.281; Table 43.)

Table 43. *Distribution of Tyneside chemical products, mid-1880s* (percentage)

Home markets	47	Netherlands	8.5
U.S.A.	13	Russian Empire	5
Germany	11.5	Others	6
Scandinavia	9		

Source: Depression of Trade, 1886, vol.2.

The time of troubles

In the early 1870s, the Tyne's chemical business, like other staple trades, went through a boom. In the four years to 1875, imports of pyrites to the river doubled. Looking forward to the use of Tees salt, the *Mining Journal* (20 Feb. 1875, p.191) was writing of the 'bright outlook': 'The unlimited command of salt on the one hand, and of cheap fuel on the other . . . will give the Tyneside alkali trade a better hold on the markets of northern Europe, so that, instead of now exporting as we do a large quantity of coal and salt for the alkali industries of other countries, we may hereafter extend our exportation of the finished articles of commerce.' Eighteen months later, the *Colliery Guardian* (30 June 1876, p.1029) still saw chemicals ranking with coal and iron as one of the three staples of the river's economy. For some time it seemed that underemployment in chemicals was just part of the general trade depression which had caused the region's wrought-iron trade even deeper distress. Within a few years, however, most other trades revived. Shipbuilding was soon expanding and in the metallurgical sector there was large-scale investment in steel works. In chemicals there was never a true recovery, but instead a few years of oscillation of trade and then a protracted decline. By the mid-1880s it was noted that every year after 1875 the chemical trade had been depressed, whereas, apart from the last two years, engineering and shipbuilding had been doing well (Depression of Trade, 1886, vol.2).

At the end of 1876 the northern correspondent of the *Mining Journal* reviewed the Tyneside industry. The chemical works then employed 'upwards' of 10,000 persons—in good times—and represented a fixed and floating capital of £3 million, and the processing of almost $1\frac{1}{2}$ million tons of raw materials. He stated, however:

> the last two years have been about the worst the chemical trade has ever known. Few firms have made any profits and the majority have incurred considerable loss. This state of affairs has had the effect of checking the plans that were projected three or four years ago for the extension of chemical works on the Tyne, but Tyneside still continues to maintain the position of the most important centre of the alkali trade in the world.

Two weeks later the journal observed, 'Like some other staple trades, it seems as if the manufacture of chemicals on the Tyne has been overdone.' (*M.J.*, 2 Dec. 1876, p.1323; 16 Dec. 1876, p.1379.) In the next four years or so the trade—and still more the expectations from it—went up and down, from a state described as giving 'more encouragement' to 'very quiet', from 'very brisk' to 'again in a most wretched state'. It was helped by the near approach of a shipping season, by a strike of Lancashire miners which hit the Merseyside alkali firms early in 1881, or badly cut down by the war in the East in the spring of 1877 (the Russo-Turkish conflict). This was in an area where the Tyne had done 'a great business'. (*M.J.*, various reports on trade.) For much of this period demand was good, but there was over-supply, with some firms unwilling to act to hold prices. By mid-1881, when business had shrunk almost to nil, every ton of soda ash sold was losing from 10*s*. to 12*s*. 6*d*.

In spite of all this uncertainty, unprofitability, and slowdown, accompanied as it was by the closure of many works, for a time the long-term production trend still seemed upward. Exports from the Tyneside works in 1875 were 30 per cent higher than in 1870, and 1880 represented a further 37 per cent increase over 1875. In 1883 Tyne works decomposed 181,000 tons of salt, a record and just twice the level of twenty years before. Thereafter, uncertainty and a loss of direction became a very definite decline.

Tyne exports fell by 10.2 per cent from 1880 to 1885. The salt consumption of 1883 proved to have marked the grand climacteric of the Tyneside alkali trade; by 1886 it had fallen by 31 per cent. (*C.T.J.*, 3 Oct. 1891, pp.207–8.) Tyneside consumed 42 per cent of the salt decomposed in British alkali works in 1874, in 1885 24 per cent, and five years later only 21 per cent. In short, the area was undergoing an absolute decline after 1883, and also, and over a longer period, a relative decline within the national Leblanc industry.

Tyneside's position within United Kingdom Leblanc alkali production

Fuel economy hit the district with cheapest fuel, and it was taking effect with great rapidity. Walter Weldon (1883) noted that from 1872 to 1882 one of the leading Leblanc firms had cut its coal consumption per ton of finished product from 3.36 to 2.16 tons. The limestone advantage had been slipping for twenty-five years. The state of the local market, as noted above, was also unfavourable; so too was that of overseas markets.

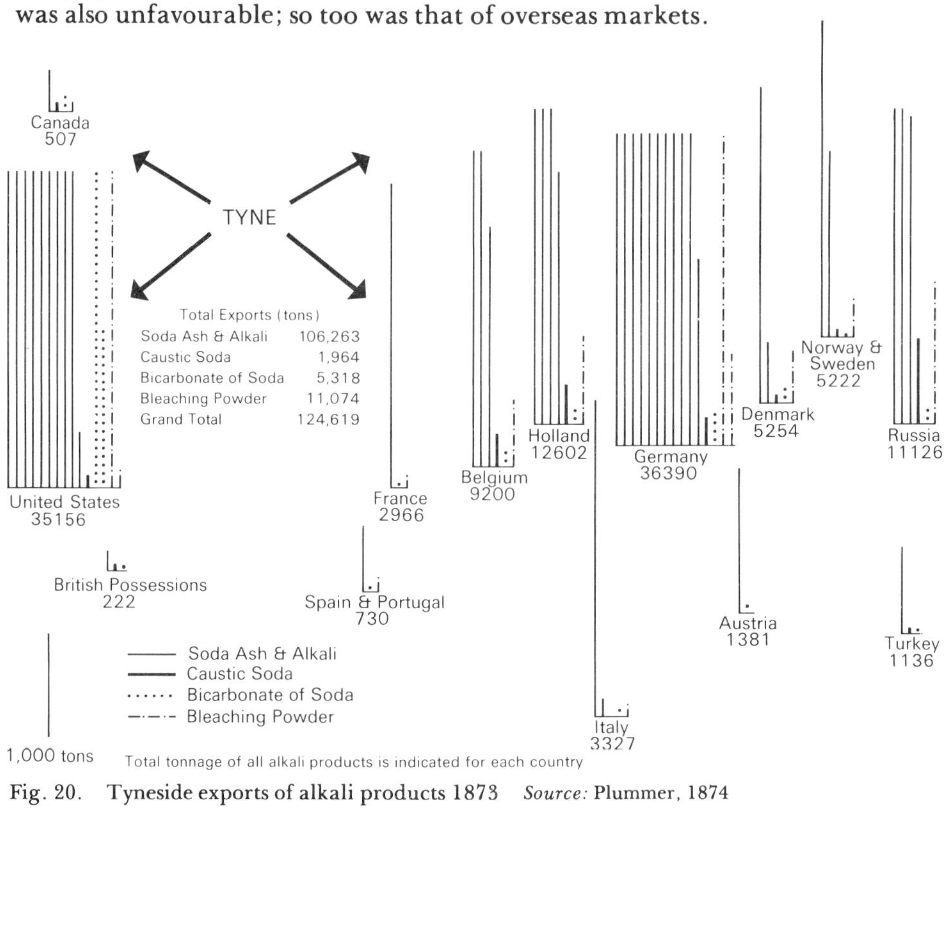

Fig. 20. Tyneside exports of alkali products 1873 *Source:* Plummer, 1874

A major overseas outlet for Tyneside chemicals lay in the countries of the North Sea Basin (Fig.20). At mid-century there had been a good deal of concern on the Tyne about French and Prussian tariff policies, and the Crimean War had hit local firms hard. Until the Cobden Treaty of 1860, British alkali was effectively shut out of France. (*M.J.*, 7 Apr. 1849; 24 June 1854, p.420; 28 Jan. 1860, p.59.) In the late 1870s the situation deteriorated rapidly. In 1879 a new Tariff Act gave effective protection to the then small industry in Germany, and within three years production there almost trebled. Exports to Germany shrank even more quickly. Similar tariff policies in the Russian Empire also hit British trade. Meanwhile, until the 1890s British alkali exports to the United States held up better. Tyneside, so dependent on the European market, suffered particularly (Tables 44 and 45).

Table 44. *Selected chemical shipments from the Tyne (coastal and foreign) 1883–1887* (thousand cwt)

		Total	*To all foreign ports*	*To Germany*
Soda ash (calcined)	1883	1539	1287	297
	1887	702	600	84
Crystal soda	1883	2382	1081	281
	1887	1986	623	39
Bleaching pwder	1883	629	462	113
	1887	594	429	68

Source: *Deutches Handel Archiv*, quoted *C.T.J.*, 3 Nov. 1888, p.271.

Table 45. *British alkali exports, 1870–1875, by value and tonnage*

	Total		*To U.S.A.*		*To others*	
	Tonnage (thousands)	Value £000	Tonnage (thousands)	Value £000	Tonnage (thousands)	Value £000
1870	193	1486	95	756	97	730
1875	251	2300	102	981	149	1319
1880	344	2398	165	1205	180	1193
1885	309	1815	167	1026	142	789
1890	317	2089	200	1257	117	832
1895	312	1558	175	770	136	788

Source: Haber, 1958, p.214.

Another disadvantage sprang from the district's specialization within the alkali trade. Soda crystals had long been a major portion of its business and much more important than in Lancashire. Conversely, caustic and bleaching powder were much less important on the Tyne. In 1866 the Tyne branch produced 86,000 tons of crystals and Lancashire only 25,000 tons, but soda crystals were the least valuable of the main products of the industry—£5. 10*s*. a ton compared with £11 for soda ash. (Kingzett, 1877, p.117.) Hydrochloric acid which in Lancashire went into bleaching powder was, for long, largely run to waste in the North-East, though eventually more bleaching powder was made. Demand for caustic soda grew in the seventies and eighties as it was used not only in soap and in paper manufacture but also in such new industries as oil refining and the manufacture of aniline dyes. In Lancashire caustic soda production went up from 1866 to 1885 from 11.2 to 140.8 thousand tons, but on the Tyne only from 3.7 to 4.1 thousand tons. In 1878 Lancashire produced 94 thousand of the 96 thousand tons of caustic soda produced in the United Kingdom. (Muspratt, 1886, pp.407–9.) As demand for these products held up well, the Tyne had to shift its emphasis to the expanding lines. Exports of soda ash in 1883 were only 1.8 per cent above those of 1873, but exports of soda crystals were 25.7 per cent and of bleaching powder 108.1 per cent up. Tyneside continued to export to the world, though tonnages fell (Fig.21).

As marketing prospects darkened and foreign competition mounted, so too did competition at home. From 1879 onwards Solvay ash production was increasingly rapidly. The Leblanc firms which survived were those which improved operations and in so doing speeded the exit of the laggards. Up to the mid-1870s Tyneside had some twenty-five alkali works. By 1883 twenty works in England had already ceased working, largely as a result of competition from the ammonia soda process. Four were in Widnes, two in St. Helens, and one in Manchester, but twelve were on Tyneside. Eight of these had already been dismantled, including the district's pioneer works at Walker. (*J.S.C.I.*, 29 Sept. 1884, p.480.)

The advance of the Solvay process

Competititon from the Brunner Mond works was discussed in Newcastle chemical circles from an early date, the very first reports causing concern there (Davidson, 1923, p.53). As early as December 1870, when expansion seemed very much the dominant theme, John Glover spoke in rather mysterious terms to the Newcastle-upon-Tyne Chemical Society, terms which antedated by over three years the similar comments of Vincent to the Society of Arts. (Vincent, 1874.) The alkali process, he remarked, still awaited the coming man, not, he thought, to introduce a shorter process than that of Leblanc but to perfect it. Almost two years later Georg Lunge noted an improvement in techniques but no breakthrough: 'Somehow or other the great revolution has not yet been accomplished. The old plant is still in existence everywhere, only augmented here and there by an appendix according to Weldon or Deacon.' (Proctor, 1883, p.397.) In February 1873, in his presidential address to the society, he observed that new ammonia soda works were being built, one of them designed for a production of almost 100 tons of soda ash weekly, and he went on:

> If the ammonia soda process really should be the process of the future, it is quite evident that those localities will have the best chance for it where strong brine is to be had for next to nothing, and that the manufacture of alkali will shift to different localities from the present. I do not mean to frighten any of my hearers with this hint, but only to exhort them to keep a sharp lookout for what is doing elsewhere, and by using every care and by adopting every improvement in the present process, which has so far outlived any other proposed, to make it able to compete with any other process in future as well. (Proctor, 1883; *T.N.C.S.* II, 1871–4.)

Fig. 21. Tyneside exports of alkali 1889 *Source: C.T.J.*, 1889

In 1874 John Pattinson again reviewed the previous few years. He commented on the secrecy of Brunner Mond's processes—'nothing is known of them outside their works'—and he quoted Weldon on their expense and slowness. The next year he was just as grudging, noting that the ammonia soda process 'is now carried on, somewhat extensively I believe, by Messrs. Brunner and Mond'. They made excellent soda, but 'it is said that the plant is very liable to get out of order, and that the alkali is produced at very high cost . . . moreover it is said that five percent of the ammonia is lost for every ton of alkali made.' (Proctor, 1883, *passim*; *T.N.C.S.* II and III.) Inaccurate information of this sort, some of it based on little more than hearsay, encouraged over-optimistic predictions of failure for the Northwich operations. As late as 1878, R. C. Clapham, then president of the society, though forced to a more realistic conclusion about the Solvay process, and to hint again at locational shifts for the industry, still clutched at straws. 'So far as can be learned of this new process, which may be said to be still in its infancy, it appears as if it might prove an important rival to the old method of working, and its practical application would point to the probable future of the soda trade as being near to the salt brines', but he speculated whether Edison's experiments with electric lighting—only under way that very year—might so cut down the demand for gas that the new works would not get their ammonia at a competitive price. At the time Brunner Mond & Co. made about 12,000 tons of alkali a year; the Tyne output of all soda products in 1877 had been 246,000 tons (*C.N.*, 8 Nov. 1878, pp.230–1.) By 1880, speaking in Newcastle, the German alkali expert R. Hasenclever suggested that the Leblanc and Solvay processes were now on equal terms. Within three years, looking back to this address, and recalling the slow emergence on Tyneside of a realistic perception of the threat from the new process, Proctor was delivering an elegy for the old method of production. 'Since that time, the Leblanc process has had a desperate struggle for life. At any rate in our neighbourhood this is evidenced by the numerous ruined castles of that industry which stud the Northumbrian shore of the Tyne.' (Proctor, 1883, p.298.) The following year Walter Weldon put the cost of production of soda ash by the Leblanc process as 70 per cent more than by Solvay's. (Weldon, 1884, p.852). In the eighties, as many Leblanc firms went to the wall and survivors had a hard time, Brunner Mond was making handsome profits—£163,000 in 1886 and £284,000 in 1890 (Reader, 1970, p.101).

Tyneside's experience in the 1870s not only illustrates an extraordinary slowness in appreciating new conditions of trade and fresh requirements in plant and location, it also shows that even when this appreciation was achieved, action did not necessarily follow. A major salt-field was known only some thirty miles south of Tyneside: no plant was built there, and there seems, moreover, to have been no speculation in all this period that Tyneside firms might contemplate such an action. Meanwhile, Leblanc Tyneside turned to various palliatives rather than radical solutions.

Labour and machinery

A variety of explanations of the difficulties of Tyneside were put forward. General depression, over-production, competition from new processes on the continent and in Britain were all considered (*M.J.*, 24 Dec. 1881, p.1583). There was a temptation to blame the workers, particularly their quality and their wages. About 1870 the first union in the chemical industry was formed on Tyneside and trouble with workmen began to mount. Early in 1873, commenting on Lunge's address (*T.N.C.S.* II, 1871–4, p.173), John Glover, who had himself entered the trade from very humble circumstances, spoke of the lower moral tone of the nation and the unfortunate attitude of the workers, and forecast a time when 'if things go on as at present upon the Tyne we shall have to play only second fiddle in the manufacture of chemicals'. Two years later Jones and Walsh of Middlesbrough were driven to mechanize their processes in part because of 'the misconduct of that very low class of workmen the decomposing men' (ibid. III, 28 Oct. 1875, p.198). By 1886 Christian Allhusen was remarking that trade unions and their strikes prevented the wage reductions he felt necessary if Britain was to compete. Wages were lower in Germany, whose trade had been lost, though he did not think that wages in Britain would have to be lowered to the Continental level (Allhusen, 1886, pp.264–5). Sometimes Tyneside men proved willing to accept wage reductions, but usually they resisted (*M.J.*, 16 Nov. 1878, p.1267). Early in 1876, when the owners tried to secure a 15 per cent cut in the men's wages, a compromise had to be reached under which a 7½ per cent reduction all round was agreed. Not all the men accepted this, and the chemical firms resorted to a lock-out and almost 6,000 were soon idle (*M.J.*, 19 Feb. 1876, p.200). In 1880 conditions were again difficult. In the autumn, the Newcastle Chemical Company tried a new way round the labour problem through a long-delayed improvement in conditions of employment; they introduced a fund to compensate all men who could not work owing to accident. (If an employee was killed in an accident the firm and the men were to contribute equally to compensation payments.) (*M.J.*, 25 Dec. 1880, p.1494.)

It has been suggested that the Factory and Workshop Act of 1878 which made new safety measures obligatory injured the industry, notably the smaller firms which lacked the necessary financial resources and had operated on a much narrower margin than the bigger firms (Campbell, 1971, pp.46–7). However Clapham (1938, p.430) pointed out that the 1878 Act made no special provision for particularly dangerous or unwholesome operations like those in parts of the Leblanc industry.

In 1885 an Industrial Remuneration Conference was held in London. The evidence submitted by Lowthian Bell and by J. Burnett of the Trades Union Congress Parliamentary Committee proved most revealing on conditions in the alkali trade (Bell & Burnett, 1885; *M.J.*, 6 Feb. 1875, pp.135–6). Bell stated that in twenty-five years the average price of products of the Tyneside industry had fallen 40 per cent, but average wages had increased by 37½ per cent.

Maintaining, with very questionable logic, that 'the new ammonia process . . . is too new and too limited in extent to have materially affected the general trade', he seemed to attribute many of the difficulties to increases in wages. Other activities on the river were prosperous and men in them were getting good wages; in short, these new or buoyant trades constituted a force which helped keep chemical wages up. Burnett reported that he and Bell had been associated in helping to settle a dispute between the Newcastle Chemical Company and its workmen ten years earlier. The outcome of that mediation penalized the smaller firms on the river. The Newcastle Chemical Company had adopted 'all the improvements in the manufacture of soda', whereas some other firms still pursued the old methods. One result of arbitration had been a cut of 10 per cent in Newcastle Chemical Company wages, but another was agreement to adjust the pay principles in all the Tyneside works. 'In the readjustment of wages in those concerns some of the workmen were found to be so underpaid that an advance of 50 percent had to be made to bring them to their proper position. That accounted to a large extent for the 35 percent increase referred to by Mr. Bell. It also explained the disappearance of those small and old-fashioned firms from the field of chemical operations in the North of England.' (Bell & Burnett, 1885, p.167.) The small firms of the region were penalized in other ways as well, notably by the mechanization of the industry and by policies designed to restrict production in a desperate effort to keep up prices.

Until economy in labour costs became important, mechanization in alkali works had not been extended far beyond the machinery necessary for grinding or for moving material (Bell, 1881, p.445). Affleck (1877) and Clapham (1878) had reviewed the advantages of mechanization at a time when both wage levels and fuel prices were low. Under these circumstances, machinery could be a doubtful blessing commercially, for wage levels rose and fell and capital tied up in plant could become a heavy burden. This may be seen in the case of the revolving ball furnaces introduced to replace hand-balling in black ash machinery (Table 46). But changes made in the industry in the twenty-five years between the first two visits of the British Association to Newcastle in 1838 and in 1863 had already involved important fuel and labour-saving innovations. (Stevenson, Clapham, & Richardson, 1863, pp.124–9.) In 1877 Affleck put the wages cost of the old labour-intensive saltcake processes at 8*s*. a ton, but with the new Jones and Walsh mechanical furnace it was no more than 6*s*. at most—though in this instance the mechanical process was over-enthusiastically adopted on the Tyne and there was some subsequent retreat. The bigger firms adopted the improvements. By 1884 Tennants of Hebburn, employing 700 men, had got rid of all hand labour in the saltcake department. (*J.S.C.I.*, 29 July 1884, pp.396–7; *M.J.*, 19 July 1884, p.852.) Newcastle Chemicals, as always, invested assiduously in all proved advances. Meanwhile the smaller works had poor machinery and could not afford to install the improvements. (Stuart, 1889, pp.145–6.) The twelve works which had been closed by 1882 had decomposed 67,000 tons of salt annually; the thirteen survivors were capable of decomposing as much as 220,000 tons (*J.S.C.I.*, 29 Sept. 1882, pp.346–7). Over a longer period it had conspicuously been the bigger works which survived (Tables 47 and 48).

There is some evidence that in machinery and techniques the Tyne was inferior to Lancashire. It was suggested in discussion that Mactear's results were certainly not startling compared with those obtained in the North-West

Table 46. *Differences between revolving and hand-operated balling furnaces at the Tyne Chemical Works, 1869*

	Hand-operated		*Revolving furnaces*
Number	25		2
Capital cost	£500		£2,000
Throughput in working week (including repair time)	21 tons		66 tons
Charging costs	—	equal	—
Fuel use per ton of saltcake charged	—	equal	—
Labour cost per ton	2*s* 6*d*.		1*s*. 6*d*.

Source: *Eng.*, 13 Aug. 1869, p.114.

Table 47. *Number of Tyneside works and tonnage of salt decomposed, 1863, 1880, 1887*

	1863	*1880*	*1887*
Number of works	18	20–5	10
Salt tonnage (thousands)	90	200	*c*. 120

Source: Stuart, 1889, p.142.

Table 48. *Changes in Leblanc plant in Tyneside area 1864–1882*

1864 (rank order) (Tonnage of salt decomposed weekly)	*Changes 1864–72*	*Changes 1872–82*	*1882* (order as in 1864) (Tonnage of salt decomposed weekly)
340			490
310			1000
263			367
240			430
116		Eliminated	—
115			149
115			148
86		Eliminated	—
86		Eliminated	—
70		Eliminated	—
43			often idle
36			340
36		Eliminated	—
36			130
36		Eliminated	—
36		Eliminated	—

1864 (rank order) (Tonnage of salt decomposed weekly)	*Changes 1864–72*	*Changes 1872–82*	*1882* (order as in 1864) (Tonnage of salt decomposed weekly)
36		Eliminated	—
36		Eliminated	—
6			125
	New plant		125
	New plant		125
	New plant		95
	New plant		muriate of potash
	New plant	Eliminated	
	New plant	Eliminated	
	New plant	Eliminated	
	New plant	Eliminated	
	New plant	Eliminated	

Source: *Annual Returns of the Alkali Inspector*, *B.P.P.*

(Mactear, 1881, pp.26–8). In 1874 John Morrison, who had until recently worked in a Lancashire caustic soda plant, reckoned that poor equipment, very wasteful of fuel and labour, largely explained Tyneside's insignificance in that line compared with Merseyside (Morrison, 1874; 1875a, pp.87–8; 1875b, pp.121, 145). Lancashire was again much more efficient in bleaching powder. In 1874, discussing the differences between some of his figures and those given by R. C. Clapham from Tyneside, Weldon (1874, p.672) observed 'somehow or other the Lancashire manufacturer carried out this process with not quite one-third the quantity of coal employed at Newcastle and at a very much less expense for wages . . .'.

Restriction of make

In the 1880s various attempts were made to reduce production and keep up prices. Generally speaking, the small firms were squeezed out by their high costs and poor bargaining power. It is true that in late 1881 and early 1882, when a 25 per cent reduction in make was being negotiated, closure of part of Allhusens and of a large section at Tennants was planned, but there is evidence of resistance to this on the part of the big firms. (*M.J.*, 1 Oct. 1881, p.1212; 22 Oct. 1881, p.1308; 10 Dec. 1881, p.1518; 7 Jan. 1882, p.26.) In May 1883 all Tyne firms were pledged to cut bleaching powder production 10 per cent (*M.J.*, 28 Apr. 1883, p.500). In terms of tonnage this would hit the big producers most, but commercially the impact was probably greater on smaller units. The case of soda crystals is even clearer. In 1884 reduction in make was agreed, on a Tyneside basis, as with bleaching powder, rather than a national basis. This did not cause much increase in price, but some economy was secured in operations. In early autumn 1886 it was decided at a meeting of the local works that the large make of soda crystals should cease at one Tyneside works. Two months later the Newcastle Chemical Company was considering making a large extension for soda crystals. (*Eng.*, 18 July 1884, p.64; *M.J.*, 9 Oct. 1886, p.1173, 4 Dec. 1886, p.1422.) Between 1878 and 1888 the number of crystal-makers on the river was cut from twelve to five, and one of the survivors was to go out of production before the end of that year 'by arrangement' (*C.T.J.*, 22 Dec. 1884, p.389). Restriction of make was not always successful in raising prices. Between December 1884 and the end of 1887, association of firms and reduced production in the Bleaching Powder Association raised prices from £6. 5*s*. to £7. 8*s*. 9*d*. a ton; over the same period the price for crystals fell from £2. 8*s*. 6*d*. to £2 a ton (*M.J.*, 31 Dec. 1887, p.1589).

Teesside salt

Discovery of salt beneath the Teesside flats was announced in the summer of 1863; production began nineteen years later. The considerable development costs made control of Teesside salt-fields possible only for the bigger firms and thus again improved their position *vis-à-vis* their smaller rivals. Up to early 1884 Newcastle Chemical Company had spent £5,480 on boring for salt without success. Two and a half years later the cost of bores was put at £6,000–£7,000 and some had already collapsed, representing a write-off of the investment. (*M.J.*, 16 Feb. 1884, p.191; 4 Sept. 1886, p.1029.) A few years later two other big firms, the Jarrow Chemical Company and Tennants, which had long co-operated closely, agreed to develop Tees salt production jointly. The smaller firms remained dependent upon purchase from their larger rivals, from independent Teesside salt producers, or from Cheshire. (*C.T.J.*, 13 Oct. 1882, p.225.)

The carrying of losses, business organization, and attitudes towards innovation

The bigger units could carry losses for years of particularly bad trade, whereas small firms went under. In 1881 Newcastle Chemical, with £0.45 million in capital, lost £15,000. A number of small works were suspended under the same trading conditions. (*J.S.C.I.*, I, 1882, p.347.) In some instances the bigger units were effectively subsidized. Much later, one manager wrote of Hebburn and Gateshead works which he had known since 1867: 'The former was maintained in operation by the courage and determination, and the purse, of the late Sir Charles Tennant Bart., and the latter similarly by the late Mr. Christian Allhusen.' (Davidson, 1923, p.50.)

At an early stage Tyneside was certainly productive of new techniques and innovations. Already in 1844, Felling works had a so-called chemical laboratory and a chemistry library (*Penny Magazine*, 25 May 1844, p.208). The revolving ball furnace was invented on the Tyne, and the Tyneside works adopted the Mactear furnace with alacrity. Yet Tyneside firms seem always to have been lacking in wide organization in material supply. It is true that, to the end of an independent alkali industry, Allhusens was one of the outstanding organizations in the country. At the meeting of shareholders in

Newcastle, called in 1890 to approve the sale of the works to the United Alkali Company, Davidson reported that, with two important exceptions, he had visited all the Leblanc works coming into the new group and found Allhusens' the best equipped and biggest of them all. Appropriately, the price paid for it was more than for any other United Alkali property. Yet in 1889 Allhusens made a profit of only £7,049. (*C.T.J.*, 22 Feb. 1890, p.130; 13 Dec. 1890.) A disturbing feature of the Tyneside trade long before its 1883 peak of over-all production was the slowness with which Teesside salt was exploited. There were some valid geological reasons, but a marked lack of enterprise is also indicated. In 1888 John Pattinson and T. W. Stuart both reflected on the lack of enterprise of their fellow Tynesiders. 'It had always seemed to him', Pattinson remarked, 'a strange thing that so many years should have been allowed to elapse between the discovery of the salt deposits and their utilization on a commercial scale. It was not wonderful that Messrs. Bolckow and Vaughan, who were iron masters, should not have pressed the thing very eagerly, but that the Tyne alkali makers should not have taken it up before seemed very strange.' Stuart added that 'the same wonder had often been in his mind'. (*J.S.C.I.*, 31 Oct. 1888, p.667, reported discussion.)

Decline and adjustment

By the late 1880s, the trade journals were wallowing in social Darwinist metaphors as they described the troubles of the alkali trade. In 1887 it was reported that 'a struggle for the survival of the fittest has been going on for several years, and thus the number of alkali works has been brought down by half in a very short time'. The *Chemical Trade Journal* (7 Jan. 1888) noted in an editorial that 'Mr. John Stuart Mill has told us that with advancing civilization profits from trade become reduced to a minimum, and if our degree of civilization be measured by the amount of profit made in the ordinary course of free trade in 1887 it may be reckoned to have reached its zenith.' In 1888 it was said 'one by one the weaker elements have gone to the wall' (*M.J.*, 5 Jan. 1889, p.10). In October the *Chemical Trade Journal* (13 Oct. 1888, p.221) had a leader entitled 'Lex Talionis, or the Alkali Maker's ABC' and the next spring returned to the advocacy of cut-throat competition: 'Let them . . . go back to the good old way of every man for himself in this world-wide trade. The fittest will survive and only the best deserve to live.' (*C.T.J.*, 13 Apr. 1889, pp.225–6.) From 25 in 1875, the number of alkali works in the North-East had fallen to 12 in 1883, and only 7 four years later (Fig.22). Trade conditions were becoming increasingly competitive, even Lancashire Leblanc works had only narrow profit margins compared with the looming competition and booming profits of the ammonia soda trade. Declining production, as indicated by the tonnage of salt decomposed, though less catastrophic than the shrinkage in the number of works, was none the less impressive.

Tyneside adjusted to changing circumstances, but with considerable

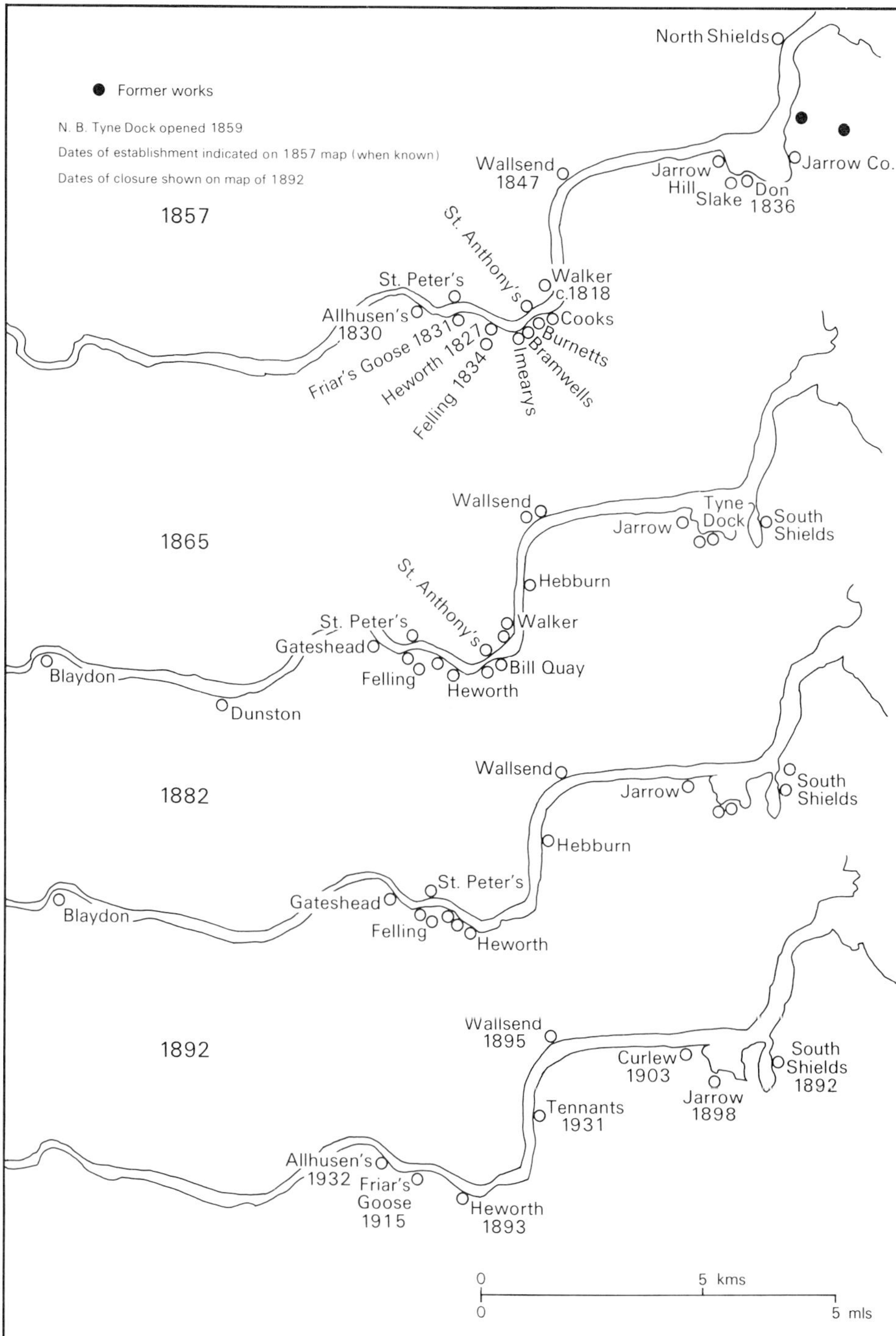

Fig. 22. Tyneside chemical works 1857, 1862, 1882, 1892 *Sources:* various

difficulty. Looking back over a dozen years or so, Stuart (1889, p.142) noted that whereas formerly hydrochloric acid had been a by-product, 'now, however, the conditions of the manufacture have so entirely changed that hydrochloric acid for the production of bleaching powder and chlorate of potash has become the object of the process while the alkalis are by-products and sold at a loss'. In 1888 the Newcastle Chemical Works spent some £23,000 on a caustic soda plant which enabled them to convert into caustic about nine-tenths of the sulphate of soda necessarily produced in association with their bleaching powder output and therefore eased their way out of soda ash (*C.T.J.*, 16 Dec. 1889). There was a complication in that bleaching powder production also involved a large output of soda ash or crystals, and therefore a still further undercutting of their prices. Over-production of both bleaching powder and soda ash was a continuing problem. Even though closure of works and the operations of the Bleaching Powder Association (1883–December 1889) helped, dependence on bleaching power remained hazardous, as was shown by the sharp drop in price on the dissolution of the Bleaching Powder Association. The low ebb and the uncertainty of the trade were well brought out in an article in the *Newcastle Chronicle* entitled 'Chemical Manufactures in the North' (reproduced *C.T.J.*, 3 Oct. 1891, pp.207–8): 'So long as the Leblanc chemical works have a monopoly of the production of bleaching powder will there be a future for the chemical trade of the Tyne, and in that life—if uncertain in duration—the Micawbers of the trade may find something turn up that will benefit them, just as the introduction of the sulphur recovery process a year or two ago has done.'

In the eighties and early nineties famous Tyneside works were dismantled, closed, or converted to new products. Carville went in 1882; Felling, with a loss of 1,400 jobs, in 1886 (*M.J.*, 13 Feb. 1886, p.193; Campbell, 1962, p.26). Allens Wallsend works closed in 1890 (*C.T.J.*, 11 Jan. 1890, pp.26, 32). The United Alkali Company took over nine Tyneside plants in autumn 1890; in the next year it closed five of them. In other cases firms changed their product line, or else the plants were bought and turned over to new products. Thus in 1891 the old Don Alkali works on Jarrow Slake was bought by the Hedworth Barium Company to produce barium dioxide, an activity which kept it at work until the early 1920s (Pilbin, 1935, p.312).

Frequently expansion of other Tyneside trades eased redeployment of labour and brought the re-use of the old chemical site as well. At Wallsend the area of the Carville works was eventually built over by Parsons turbine works, and Swan Hunter extended its shipyards over Allen's (Campbell, 1962, p.37). Early in the twentieth century United Alkali was trying to interest the Co-operative Wholesale Society and also Cammell Laird in sites at Heworth and Hebburn (U.A.C., Tyne, 10 Aug. 1891; U.A.C., Liverpool, 19 Mar. 1906, 21 Jan. 1907).

The decline may be seen clearly in the development of the river's largest works, Allhusens'. When the Newcastle Chemical Company was formed in

1873 as a public concern to take over from C. Allhusen and Company, it had a capital of £0.6 million and employed 2,500 men. Profit the previous year had been £89,000. In 1884 profits were down to £20,000 and the nominal capital was reduced from £510,000 to £240,000; at the same time the nominal capital of Brunner Mond was being increased from £0.6 million to £1.5 million. (*M.J.*, 16 Feb. 1884, p.191; Watts, 1923, p.33.) By 1888 employment at Allhusens had been cut drastically to 'over 1500' (*C.G.*, 2 Mar. 1888, p.314).

As early as 1882 when an attempt was made to sell 1,000 £10 shares in the Newcastle Chemical Company, for which £7. 10*s*. had previously been paid, no offers could be obtained. By contrast, when the engineering and shipbuilding group of Armstrong-Mitchell of Elswick and Walker was formed and wished to raise £65,000 capital, it had offers almost at once for about three times that amount. Even the Tynedale and Nenthead Lead Company, a new concern in the highly uncertain Pennine mining tracts, was quickly able to find takers for one-third of its capital issue. Chemicals had clearly ceased to be a glamour industry on Tyneside. (*M.J.*, 1 July 1882, p.799; 9 Dec. 1882, p.1503; 23 Dec. 1882, p.1551.) When Christian Allhusen died early in 1890, his shares in his old company were put at no more than £47,000, as opposed to £600,000 in other public companies, £140,500 in American railroad securities, and £76,950 in stocks and bonds of foreign governments (*C.T.J.*, 8 Mar. 1890). In December 1890 the shareholders of the Newcastle Chemical Works Company were happy to vote heavily in favour of a sale to the United Alkali Company—51,923 out of 60,000 ordinary shares and 57,767 out of 60,000 preference shares being in favour.

The decline had wide effects in the district. In 1866–8 some 6,000–8,000 men had been employed, and later a maximum of about 10,000 (Lunge, 1880, p.68; Pollution Commission, 1870). The alkali works consumed annually 0.65 million tons of coal, all locally raised. Of the 20,000 vessels which then came into the Tyne each year, about one-quarter discharged materials for the alkali works (Pollution Commission, 1870). By 1893 employment in the Tyneside United Alkali works was only 2,900, a little under half that of its works in Widnes alone, though Allhusens' workforce was half as big again as that in the largest single works there. The Tyneside and National Labourers Union had a total of 4,200 Tyneside chemical workers. (Labour, 1893/4.) Male employment by 'manufacturing chemists' in County Durham in 1891 totalled 2,016, or less than twice the level of 1851 (1,089); in Lancashire and Cheshire employment in 1851 had been 2,415, but by 1891 it was 12,207. By 1901 alkali manufacture employed 966 in Durham and 7,943 in the two north-western counties. (Welton, 1897/8; *B.P.P.*, 1905, LXXXIV.) Coastal shipments and exports shrank (Table 49), and with this decline other employment opportunities were lost.

Although by the mid-1890s the Tyneside alkali industry was of small account, its demise was to be protracted. By 1898 there were only four works.

Table 49. *Exports and coastal shipments of alkali from the Tyne* (thousand tons)

	1859	*1873*	*1883*	*1887*	*1892*	*1898*
Soda ash	13.5	63.2	64.4	35.1	10.6	8.1
Soda crystals	15.0	43.0	54.0	99.5	54.6	17.6
Bicarbonate of soda	2.3	N.A.	N.A.	2.3	—	—
Bleaching powder	2.9	11.1	23.1	29.7	28.7	23.3
Caustic soda	N.A.	N.A.	N.A.	4.5	20.3	12.8
Total	*c.* 34[a]	117[a]	141[a]	171	114	55

[a] Total of items listed only.

Sources: various.

Richardsons at Jarrow closed in 1903, Friars Goose in 1915. Tennants' Hebburn works survived until 1931 and Allhusens' to 1932, but operations had been emasculated long before that, the Leblanc process ceasing at Hebburn in 1900 and at Allhusens in October 1920. In 1932, when the Gateshead works closed, some of its production functions were transferred to Billingham. (Stuart, 1932; I.C.I., List of United Alkali Company Works.)

In 1901 the Tyneside correspondent of *Engineering* (12 July 1901, pp.54–5) was depressed: Tyneside, 'at one time so important a chemical centre, has, of late years, in the dismantled works which line the river's banks become a standing monument to the uncertainty which ever attends chemical enterprise. Like Marius musing over the ruins of Carthage, we have, when gazing on the relics of the chemical industry of Tyneside, been unable to avoid a feeling of pessimism.' The period of rapid contraction after 1890, however, was not without its own interest: it saw the establishing of the pre-conditions for the growth of a major new chemical industry a generation later and thirty miles away to the south, on Teesside.

CHAPTER 12

Salt and Alkali on Teesside

As early as the 1860s, when it was the world's most rapidly growing iron-making centre, there were already hopes that Teesside might also become a leading focus of the alkali trade. These early ambitions were frustrated; more modest ones, of a vital link with Tyneside chemical manufacture, though achieved, proved disappointing in their implications for Teesside economic growth.

The discovery and exploitation of Teesside salt

As the first ironworks came into production, in the early 1850s the possibility of salt lying below the Tees estuary was raised by Greenwell in a work on marine engineering; there were even hints of it in Young's *Studies of the Geology of the Yorkshire Coast* published in 1828 (Peacock, 1869, p.336). In July 1859 the leading iron firm, Bolckow and Vaughan, began to bore for water. By August 1862 they had proved a bed of salt 100–20 ft thick. These facts are well documented but two major puzzles remain. Bolckows proved the salt at a depth of 1,200 ft. It is inconceivable that they should have bored so deeply for water; clearly, therefore, they were searching for minerals. Secondly, having discovered salt, why did they wait a year before announcing it? The answer to the first, and part of the answer to the second, was provided some thirty years later by one of Vaughan's successors as manager of Bolckow and Vaughan, E. Windsor Richards. Richards noted that in fact John Vaughan had been boring for coal, but, fearing that he would be laughed at by those familiar with the geology of the North-East, announced that he had been looking for water (*Eng.*, 11 Aug. 1893, p.166). Disappointment may account for the year's delay, but another factor was the general uncertainty of trade in 1862–3 because of the American Civil War. Certainly the lag over salt contrasted sharply with the firm's earlier alacrity in opening up the Eston iron mines. There was no long-term deficiency in demand for salt: in 1861 the North-East consumed 125,000 tons of salt in alkali manufacture, and consumption was growing rapdily. In addition there was domestic demand, a large requirement for East Coast fisheries, and export prospects in northern Europe. The explanation for the delay lay partly in the geological conditions of Teesside salt and partly in the general trade context of the Teesside area.

Three characteristics distinguishing Teesside salt deposits from those of Cheshire constitute the main physical obstacles to their development. The salt bed is deep, covered by water-bearing strata, and yet does not produce a natural brine. In Cheshire salt was often no more than one-tenth as deep

(Spons, 1879, p.310). As Teesside salt did not produce a natural brine, it was necessary either to mine rock-salt or produce an artificial brine. The great thickness of overlying strata — alluvium, clay, Bunter Sandstone, and Permian marls — made both raising the salt and lifting the brine costly; its heavily watered condition made it difficult to keep open either shafts or boreholes. There was one compensating advantage: subsidence of the surface was not a problem, but this social cost of Cheshire salt was only belatedly brought home to the firms which extracted it.

The business context of the discovery was also unfavourable to large-scale early development. The early sixties were unpropitious, the late sixties had years of still greater uncertainty. In the early 1870s Bolckows, like all other iron firms, were preoccupied with a great boom, and from 1874 the iron trade was plunged in an extended depression which eventually involved the writing-off of most of the wrought-iron works in the district. In the mid-1870s Bolckows were completing their new Bessemer works at Eston. Such conditions provided sound reasons for delay in a completely different sphere of business, and cut off what, in the four years 1869 to 1873, had seemed a promising beginning. In the year to spring 1870, Bolckow Vaughan and Co. made a profit of £150,000. They decided to devote £40,000 of this to sinking to the salt bed. (*C.G.*, 11 Apr. 1870, p.258.) Late in 1869 work on two shafts began; within two or three months a depth of 75 ft was reached, by March 1871 310 ft. Great difficulties in keeping water out had delayed progress. In March 1872 the shaft had reached a depth of 489 feet. (*Eng.*, 18 Feb. 1870, p.110, 22 Apr. 1870, p.283; *C.G.*, 17 Mar. 1871, p.283; *M.J.*, 16 Mar. 1872.) By late summer 1873, before the iron boom burst, it had already been announced that Bolckows would not proceed further at that time with their salt sinking. The journal *Engineering* (5 Sept. 1873, p.183) observed, 'It is believed that it is simply on account of their being so busily engaged in the iron trade, in connection with which they are making great extensions.' The initiative now passed to others.

In spring 1870 the Tees Salt Company was incorporated, as its prospectus put it, 'for the purpose of working rock salt, under property at Middlesbrough-on-Tees, and converting the rock salt so obtained into crystalline salt, for the purposes of the alkali and other trades'. The lease was for forty-two years, and the company claimed that it would produce 100,000 tons of salt annually (*Eng.*, 11 Mar. 1870, p.157). After a great initial flourish it soon vanished from the records.

Bell Brothers of Port Clarence were next. They approached the matter cautiously, without, apparently, the ambitions of Bolckows or Tees Salt. In 1874, testing near their ironworks with a diamond borer, they proved the salt between 1,127 and 1,222 feet. Deterred by Bolckows' experience and by the depression in iron, they postponed further development for some years, but in February 1877 it was decided to send Thomas Bell to the Continent to report on methods adopted there for working salt. (*M.J.*, 1 June 1878, p.596; Bell

Minute Books, 6 Feb. 1877, p.288.) In 1880 he suggested the use of a process employed in the Lorraine salt fields (though only at a depth of about 400 feet). This involved sinking a double tube, pumping in water through the central one in order to dissolve the rock-salt into a brine, and then pumping up the brine through the outer tube. In spring 1880 boring began at Saltholme Farm. By April/May 1881 salt was reached at 1,043 feet. Evaporating pans were installed at Port Clarence, and the first Teesside salt produced on 1 August 1882 was sold to the Tyne Alkali Company. Within six years four other concerns were making salt on Teesside (Fig.23; Tables 50 and 51).

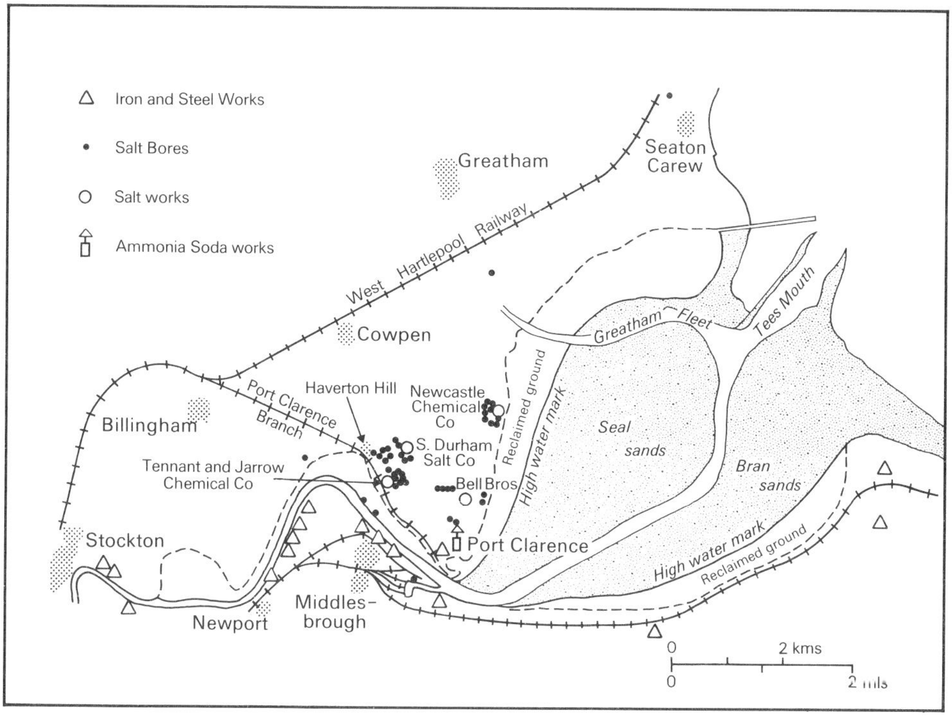

Fig. 23. The Teesside salt district 1888 Based on: *J.S.C.I.*, 31 Oct. 1888

Table 50. *Salt on Teesside, 1888*

Company	*No. wells*	*No. salt-pans*
Bell Bros.	8	19
Newcastle Chemical Works	8	22
South Durham Salt Co.	7	11
C. Tennant and Partners and the Jarrow Chemical Co.	8	20
Bolckow Vaughan & Co.	2	9
	33	81

Source: Stuart, 1888.

Table 51. *Teesside weekly salt production, 1883–1887* (tons)

Spring 1883	300
Spring 1884	500
Autumn 1887	4500

Sources: various.

Output mounted rapidly, and by 1890 Teesside produced over 200,000 tons annually. Some was table salt; fish-trade salt was an important line, and there were deliveries to the West of Scotland alkali works; but the largest outlet was on the Tyne, whose alkali works in their twilight years at last became independent of Cheshire.

Costs and charges for Teesside salt

When the discovery was announced in 1863, it was remarked significantly that almost two-thirds of the price paid for Cheshire salt on the Tyne was made up by the cost of carriage (*M.J.*, 10 Oct. 1863, p.727). There was the clear expectation that Teesside salt would be cheaper than Cheshire salt in the North-East. In 1869/70 it was reckoned that salt which was 3*s*. 9*d*. a ton on the Weaver cost from 7*s*. to 8*s*. 3*d*. to carry to the Tyne. Much Cleveland iron ore had been delivered to the Tyne coastwise for a little over 1*s*. a ton, and it was reckoned that salt carried similarly would cost no more than 2*s*. a ton to deliver. (*North Eastern Gazette*, 16 Nov. 1869; *I.C.T.R.*, 16 Mar. 1870, pp.166–7.) Twenty years later, experience justified these predictions, salt being loaded at Port Clarence jetty for $\frac{1}{2}$*d*. a ton and delivered coastwise to Gateshead for from 1*s*. 3*d*. to 1*s*. 6*d*., a saving of at least 1*s*. on the rail rate (*C.T.J.*, 16 Feb. 1889; Table 52).

Table 52. *Charges for Cheshire and Teesside salt delivered to the Tyne, 1889*

		s.	*d.*
Cheshire salt		9	6
Tees salt			
Loading costs	up to		3
Carriage (a) by rail		2	9
(b) by water		1	6

Note: Weaver salt could be landed in Liverpool for 2*s*. 3*d*. a ton carriage.

Source: Stuart, 1889, *passim*.

However, Teesside salt delivered on the Tyne did not, as had been expected, sell much below Cheshire salt. The first sales in 1882 were made either at parity with delivered prices for Cheshire salt or at no more than 9*d*. a ton less. (*M.J.*, 12 Aug. 1882, p.987; *J.S.C.I.*, I, 1882, p.407.) Sometimes the gap between the two prices was greater—in 1886, for instance, Tees salt was generally 2*s*. 6*d*. a ton cheaper (*Eng.*, 27 July 1886, p.859). Generally, however, any advantage was marginal. The main reason was that higher production costs on Teesside salt counterbalanced its lower delivery charges.

Because of the depth of the salt beds, both overheads and direct operating costs were higher. This was recognized from the start—even the generally optimistic Tees Salt Company prospectus of 1870 put the cost of salt on barges in the Tees at 5*s*. a ton, compared with 3*s*. 9*d*. for salt on the Weaver. (*I.C.T.R.*, 16 Mar. 1870, pp.166–7.) Deep bores implied high capital costs: there was greater uncertainty—the first wells sunk by the Newcastle Chemical Company near the banks of the Tees penetrated salt beds which were too thin to work; a danger of being overwhelmed by water; and the problem of the caving-in of the well-sides (*C.G.*, 11 Sept. 1885, p.416 on the collapse of one of Bell Brothers' five boreholes). The process employed in working the Teesside salt increased costs still more, though in 1889 small coal for the evaporating pans was only 4*s*. a ton on the Tees as opposed to 5*s*. a ton in Cheshire salt-works, and some Teesside operators, notably Bells, avoided even this by using blast furnace waste heat (Stuart, 1889, p.666; Bell, 1887, p.141).

Undoubtedly the three Tyneside alkali firms which took a direct part in Teesside salt gained from control of their own sources of supply; in relative terms smaller concerns on the Tyne were disadvantaged still further by the change. Even when allowance is made for the considerable technical difficulties, all the uncertainties and the increased and less predictable costs of working, the question remains why Tyneside firms allowed over twenty years to pass from salt discovery to salt exploitation. It is hard to avoid the conclusion that they showed a lamentable lack of enterprise. Certainly the opportunity was recognized by some, both inside and outside the trade. 'This discovery', wrote three Tyneside experts only a few months after it was made, 'may render the alkali trade of the district independent of Cheshire and develop a large export trade in salt.' (Armstrong, 1864, p.164.) Yet six years later when the ill-fated Tees Salt Company was formed, all its eight directors were Teesside men (*I.C.T.R.*, 9 Mar. 1870, p.155).

Teesside chemical manufacture

Before the production of salt in the district, Teesside had two alkali firms. Jones and Walsh, established in 1867, found that, in contrast to the massed ranks of Tyneside, which allowed each firm to blame others for pollution, their position on Teesside exposed them to the complaints of the public and the attentions of the alkali inspector. Their significant response, spurred on by the even greater incentive of labour troubles, was to pioneer the introduction of the first mechanical saltcake furnace in 1875 (*T.N.C.S.*, 28 Oct. 1875, p.198). Sadlers were established in Cargo Fleet Road in 1868. As soon as salt was proved, it was widely expected that, as well as boosting the Tyneside industry, it would bring more chemical production to Teesside and that Tyneside would eventually have difficulty in competing with the local industry (*M.J.*, 5 Sept. 1863, p.635; *North Eastern Gazette*, 19 Nov. 1869; see also *I.C.T.R.*, 9 Mar. 1870, p.155, 16 Mar. 1870, p.166). In fact there were difficulties. Even with local salt, Teesside was intrinsically ill suited for the full

Leblanc process. Fuel economy was marked throughout the history of the process, while consumption of salt remained more or less constant at 25 cwt. (Haber, 1958, p.100.) Yet, for the complete Leblanc process, proximity to coal remained more attractive than nearness to salt. In the middle seventies the nearest coal-mines to Teesside were South Wingate, Trimdon, and Bishop Middleham, ranging from 10 to 13 miles from Port Clarence, and at that time Jones and Walsh burned 250 tons of fuel weekly. (*T.N.C.S.*, 20 Oct. 1875, p.244.) However, if Teesside had concentrated on the production of saltcake for further processing into soda ash or caustic soda elsewhere, it might have been a more attractive location.

One of the most serious disadvantages was lack of shipping facilities to match those of the Tyne. At a meeting held in Middlesbrough in 1868 to 'memorialize' the North Eastern Railway on the need for better docks, one speaker noted how it was often necessary to ship iron from other British ports to foreign destinations while they had to make do with 'that little bit of water which they dignified by the name of dock' (*Middlesbrough and Stockton Gazette*, 5 June 1868, p.5). A few years later James Young, the petroleum expert, who had once worked as chemist for Muspratt, was contemplating manufacturing with what was described as a new process involving ammonia and carbonic acid gas (a variant of the ammonia soda process) which was reckoned to make soda ash at half to under one-third of current costs. Considerable fuel economy was involved, and Young was reported to be considering building his plant on Teesside. (*M.J.*, 6 Feb. 1875, pp.135–6.) By virtue of its much lower level of fuel consumption and the need for brine rather than salt, Solvay's process seemed ideal for Teesside, so that Ward (1883, p.254) suggested 'it may come to pass that the soda industry of the north may to a very large extent migrate from the Tyne to the banks of the Tees', where a contemporary foresaw that 'on the low-lying . . . shore between the Tees and West Hartlepool large chemical works may be early erected' (*Eng.*, 15 Sept. 1882, p.265). Speculation, however, far outran achievement.

In spring 1883, when Allhusens purchased land for salt bores near Bells' existing works, it was rumoured that they would build alkali works. Sadlers was said to have been taken over by interests with 'large capital' intending not only to put down boreholes but to operate the ammonia soda process (*M.J.*, 24 Feb. 1883, p.240; 17 Mar. 1883, p.322; 14 Apr. 1883, p.452). Some years later the formation of a new company, the Great Northern Salt and Chemical Works with capital of £200,000, was reported, with plans to acquire 36 acres at Haverton Hill to work salt and to produce soda and other chemicals (*M.J.*, 5 May 1888, p.519). As the introduction of efficient rail carriage and the building of wharfs into the Tees with coastwise shipment improved salt delivery to Tyneside, it seemed that the Tees would gain less from the mineral discovery than its northern rival. When it seemed possible that brine rather than salt might be delivered to the Tyne, local interests were aroused. Some were emphatic that brine pumping should not be allowed to points outside Teesside. 'If this were permitted, the salt district might suffer all the injury caused by subsidence, without having at present any remedy, and without deriving the advantages that would accrue from the manufacture of salt or alkali in the district.'

Port Clarence works

One ammonia soda works was built and there were fairly serious proposals for two others. Bells planned a works as early as 1881, began to build in 1883, and brought it into production in early autumn of that year, making it third only to Winnington and Sandbach among British works. Port Clarence never became an important producer of alkali. It turned out 8,000 tons of ammonia soda in 1887; two years earlier Brunner Mond & Co. had made 77,000 tons. (*M.J.*, 17 Nov. 1888, p.1301.) It was extended shortly after being brought into production, and in 1890, having sold their salt wells to the Salt Union, Bells were putting down two more bores to supply it. Even so it was still characterized as experimental and as playing no important role in the trade (British Association, 1889, p.148; *J.S.C.I.*, 30 June 1890, p.672).

The reasons for this slow and unspectacular growth are complex. They undoubtedly include Bell Brothers' continuing preoccupation with iron, and the pressing issue of when a leading iron firm should enter the steel trade. Bells had wide commitments: long-established interests in Tyneside chemicals and in Washington, where in the sixties and early seventies they had also produced aluminium. The character of Isaac Lowthian Bell, the dominant force in the firm, was another liability. Bell was an ageing industrial patriarch, whose views, even of the iron trade in which he had achieved world-wide renown, were becoming markedly conservative. Finally, Bell Brothers' problems were associated with the particular process they adopted—the ammonia soda process as in Schloesing's patents of the 1850s, not in the Solvay pattern. Evidence suggests that Schloesing's process was more costly.

Overcoming running-in difficulties, Bells cut production costs from an average of 91*s*. 5*d*. in the half-year to December 1885 to under 80*s*. a ton at the end of it. Ten years later they were only 48.16*s* a ton. (Bell, Minute Books; Clarence, 1896.) In the second half of 1887 they built the second element on to the plant and also started to make soda crystals at their Bill Quay works. By 1895, when the transfer of operations to a limited company was made, total assets were put at £978,000, of which Clarence Steelworks was £43,000 and the Clarence Alkali Works £74,000. Eventually the combination of the width of their industrial interests and their limited resources meant that they succeeded neither in chemicals nor in steel. On 1 January 1896 Brunner Mond & Co. took over Port Clarence operations, but they did not own the works until 1898. Willingness to sell was not only the result of Bells' preoccupation with steel but also a response to menacing moves from others.

Early in 1895 Brunners had been said to have almost completed negotiations for a Teesside works capable of dealing with as much as 200,000 tons of salt annually—or some sixteen times Bells' level of activity. They were expected to obtain this brine from the Salt Union and from Bells. (*C.T.J.*, 5 Jan. 1895, p.7, 18 May 1895, p.322; *J.S.C.I.*, 31 May 1895, p.527.) At the same time the United Alkali Company seemed likely to respond on Teesside. However, at the end of 1895, the United Alkali Company, Brunner Mond & Co., and Bowman Thompson came to an agreement on prices and the prospect of competition in building ammonia soda works on Teesside faded. But, as John Brunner told his General Meeting of 1897, there were good prospects in their negotiations with Bells: 'In addition to these works, they had acquired a very considerable freehold estate and very valuable salt rights. The situation had excellent shipping facilities, and was a very valuable one for the manufacture of ammonia soda.' (*C.T.J.*, 29 May 1897, p.367.)

Brunner Mond & Co. closed Port Clarence works, and shortly after 1900 demolished them. There was, however, an interesting postscript. At the beginning of World War I a Haber Bosch plant for synthetic ammonia was built at Oppau near Ludwigshafen. During the war the Badische company built a much bigger works at Merseburg in the Stassfurt salt area. Carbon dioxide was a by-product, and this, and the availability of salt, made the location ideal for soda ash production by the ammonia soda process. After British experts had studied German developments, partly by means of a blatant tour of industrial espionage in 1919, it was decided that the possible German soda ash threat should be met by locating the British synthetic ammonia plant on a salt-field. Billingham on Teesside seemed ideal: it was excellently placed near the new Billingham station of the Newcastle Electric Supply Company and lay on top of immense salt deposits. It was expected that the ammonia works would be accompanied by a large Solvay plant. However, as late as autumn 1925, T. W. Stuart, who had joined Allhusens fifty-nine years before, when writing to Brunner Mond & Co. about ammonia soda development observed that foundation conditions at the old Allhusens salt works were not good, so that 'we would not advise the Ammonia Soda works being put there'. Two years later there was still discussion of a Billingham Solvay works turning out 1,000 tons of soda ash daily, a tonnage equal to well over half Brunners' record output that year. (I.C.I. Salt.) No ammonia soda capacity was ever built at Billingham. In spite of all its advantages and its great success subsequently in a wide range of other heavy chemicals, Teesside has never achieved prominence in the alkali industry.

CHAPTER 13

United Alkali Company: formation and rationalization

Antecedents

'The plant is extremely costly and expensive, and is subject to the severest possible wear and tear, while the process is attended by an inevitable creation of nuisances which desolate its surroundings. This being so, greatly as mankind is its debtor, few would regret to see it superseded by the more cleanly, simpler and incomparably cheaper Ammonia Soda process. . . .' (Wyatt, 1893, p.58.) Not all shared Wyatt's assessment of the Leblanc process. On 1 November 1890 the United Alkali Company (U.A.C.) was formed, as a defensive amalgamation of over 90 per cent of the British Leblanc firms, and with a capital of £6 million. It was the biggest manufacturing concern in Britain. A foreign commentator saw one of the consequences: U.A.C., 'in its hurry to control the trade, loaded itself at high prices with plants now obsolete or nearly so—in the usual combination style' (*Engineering and Mining Journal*, 17 Aug. 1901). In summer 1890, as the United Alkali Company began to take shape after extensive negotiations, the editor of the *Chemical Trade Journal* (26 July 1890, pp.45–6) also perceived its significance: 'To our mind it is the last act of a great drama—an act which will depend for its duration in great measure upon the ability of the players themselves.' In the event the duration was over thirty-five years, but long before that most of the players had been changed, and, though the name remained, the products of the concern had altered almost out of recognition. The last, long act, was accompanied by the dirge of a long-drawn-out decline. When it finally merged into the new I.C.I., U.A.C., once a prince among companies, was an industrial pauper.

The genesis of the United Alkali Company can be traced to earlier attempts at combination or co-operation. In December 1838 there had been some loose form of association or understanding between Tyneside soda-makers and Tennants, Chances of Oldbury, and the British Alkali Company of Bromsgrove (Barker & Harris, 1954, pp.253, 254). By 1867 a group of alkali manufacturers had set up the Tharsis Copper and Sulphur Company to acquire Spanish mines, ensure sulphur supplies, and extract copper. In the 1870s there was a close understanding between Gaskell Deacons in Widnes and the Netham Chemical Company in Bristol. In September 1871 John Howard of Gaskell Deacons stated that 'a frank exchange of views between manufacturers is at all times desirable'. By 1882, when competititon from ammonia soda and depression of trade generally were becoming serious to the Leblanc firms. Holbrook Gaskell wrote—to whom is not clear—about the control of the trade's supplies of materials. He said that, at the next meeting of the directors of the Tharsis Company, he as a director would urge that they cut the sulphur price to allow Leblanc material to compete with the ammonia soda business. Writing to Chance in April 1882 he noted that Tharsis could afford to cut prices, for their profits were 30 per cent, and that Tennant and he as large shareholding directors 'are endeavouring to open their eyes'. (Correspondence from Gaskell Deacon and Company in I.C.I., Runcorn Records.) A more immediate precedent for amalgamation was the Lancashire Bleaching Powder Association, which widened to become the Bleaching Powder Association in 1883 and included most of the producers; it had fixed prices and allocated tonnage quotas. It also bought up and dismantled a number of works. (*C.T.J.*, 23 June 1888, p.396.) Within the B.P.A. contacts became closer, and some members advocated a complete amalgamation rather than a trade association. The B.P.A. failed however, and on its dissolution in December 1889 prices fell by £2 a ton. Another model and incentive for amalgamation was provided in 1888 by the Salt Union which claimed to control 91 per cent of United Kingdom salt output. In the last weeks of 1888 a combination was formed among Widnes alkali firms in order to secure the newly discovered Fleetwood salt-fields so that they might be able to supply their needs by water and thus break their dependence upon the Salt Union. (*C.T.J.*, 8 Dec. 1888, p.357.) The following summer attempts were made to produce a Leblanc syndicate by restricting the tonnage of salt decomposed, leaving prices and specialization to individual works, but there was deep division between the thirty-seven firms involved (Table 53: *C.T.J.*, 6 July 1889, pp.1, 2, 13 July 1889, p.14).

Table 53. *Leblanc syndicate discussions, summer 1889*

	No. of firms	*Weekly salt tonnage decomposed*
Agreed	20	7053
Conditional agreement	7	3405
Waiting	3 or 4	1448
Rejection	6	1426

Source: *C.T.J.*, 13 July 1889, p.14.

Eventually half a dozen partners of the biggest concerns formed a committee to interview both individual former members of the Bleaching Powder Association and outsiders. Dispute over sales prices to the proposed new concern delayed acceptance, even by those favourably inclined, for at least a year. (U.A.C., 1923, pp.45, 46.) The year 1889 proved a testing one, but negotiations continued nevertheless. In March 1890 John Brock, who seven years earlier had been one of the chief promoters of the Lancashire Bleaching Powder Association and was a leader of the trade in Widnes, and James H. Denning, of St. Helens and Runcorn, acting as nominees of the committee sent out a circular to firms in the trade. It stated

The undersigned have, at the request of several influential members of the Leblanc Alkali Trade, privately interviewed nearly all the members of that trade, and now beg to report, that, with very few exceptions, the trade is of the opinion that a 'Chemical Union' composed of all the Leblanc works would be most desirable provided the works can be obtained by the 'Union' at a fair, just and equitable valuation. . . .

The circular stressed, 'It is most necessary that values of works be kept as low as possible, otherwise the "Union" would be overweighted with capital, which would be fatal to its success.' (Brock and Dennis, 1890.)

Starting in September with St. Helens works, by the end of October 1890 agreements had been signed to take over forty firms mainly in the Leblanc trade, the Leblanc departments of five other concerns, and three salt properties (Fig.24). The works taken over had immense capacity. In addition to main products they already covered a range of other chemicals, treating up to 100,000 tons of copper ore to extract copper, gold, silver, producing superphosphates and other manures, soap and glycerine, pearl hardening, bichromate, carbonate and caustic potash, hyposulphite of soda, silicate of soda, manganate of soda, salammoniac, sulphate of ammonia, ultramarine, and crystal carbonate (Table 54).

Table 54. *Main products of United Alkali Company plants, 1890, 1907, and 1911* (thousand tons)

	1890 (capacity)	*1907* (production)	*1911* (capacity)
Sulphuric acid	700	N.A.	555
Salt decomposed	600	40	N.A.
Bleaching powder	150	75	101
Caustic soda	180	70	95
Soda ash	140	60	105
Soda crystals	140	140	104
Sulphate of soda	60	60	N.A.
Sulphur	40	30	30
Chlorate of potash	5	4	5
Bicarbonate of soda	N.A.	N.A.	N.A.
Hydrochloric acid	N.A.	N.A.	117

Sources: U.A.C., prospectus 1890; U.A.C., Records.

The prospectus of the new company foresaw troublesome times. It recognized that, 'Carbonates of soda are produced more cheaply by the Ammonia Soda process', but none the less anticipated satisfactory financial returns through operating profits, avoidance of undue competition, paralleled by cost economies from the unifying and concentrating of management and from better access to supply. It also promised technical improvement, profits coming from 'the increased efficiency of all the manufacturies from applying to each the high standard of working which the best experienced, and technical ability of the trade will show to be possible' (U.A.C., 1890). It was encouraging that over 85 per cent of the share capital in U.A.C. was taken up by those already in the industry. A few withdrew—Neil Mathieson insisted on having cash rather than shares, and invested in

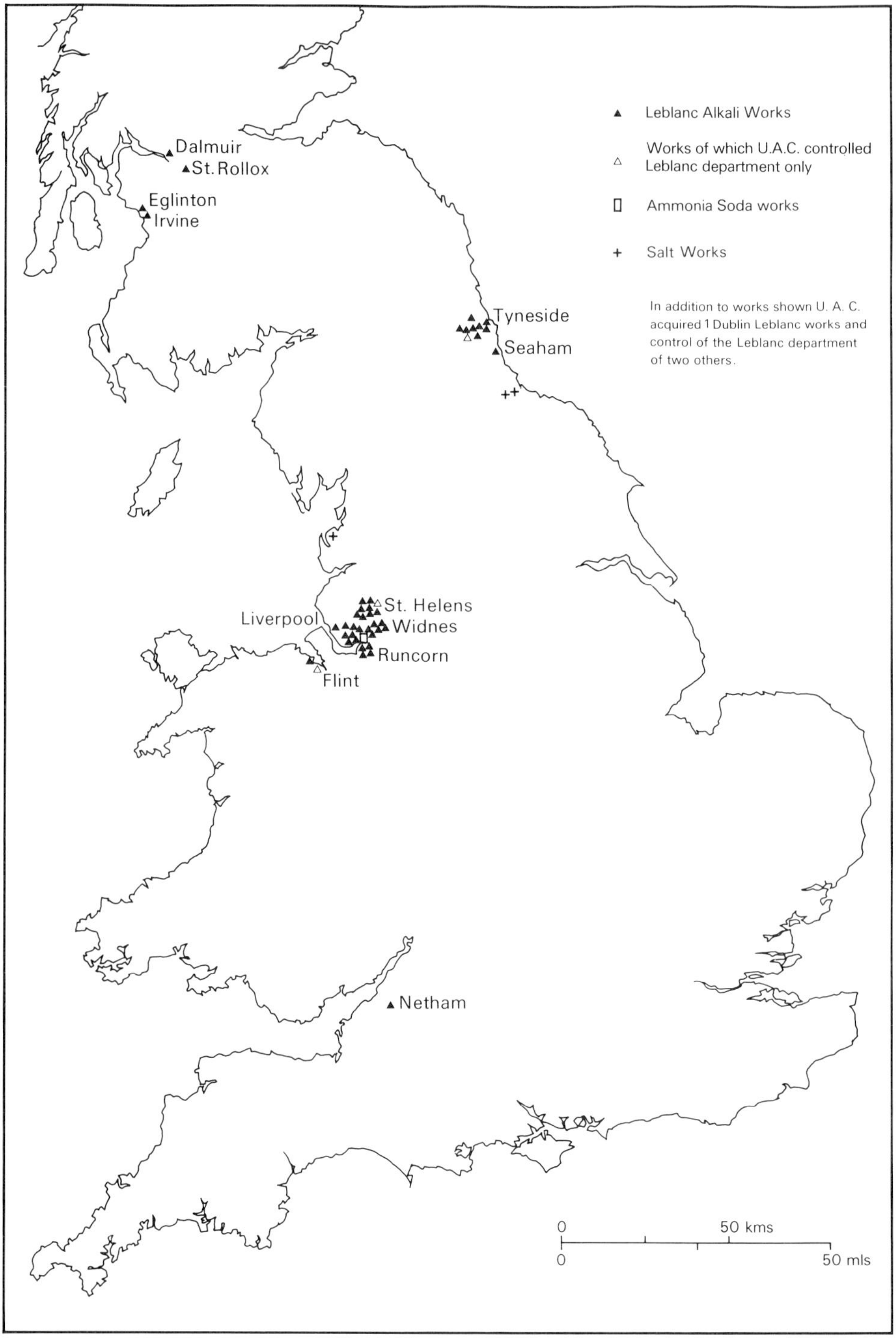

Fig. 24. The United Alkali Co. 1890 *Source:* U.A.C., 1890

United States chemical expansion. It quickly became clear that plant specialization and the elimination of unnecessary haulage were not the complete panacea; capacity was so inflated that some plants would have to close. Thus, though the works had capacity to decompose 600,000 tons of salt, in no year after 1892 did the combined British Leblanc works process more than 500,000 tons.

U.A.C. had first to convince the commercial world that it would be efficient. In September 1890, at a critical stage in negotiations, Charles Tennant, J. C. Stevenson, E. K. Muspratt, and others wrote to *The Times* to defend their emerging company against the charge that it was putting together a collection of 'old' or 'obsolete' works. The Leblanc trade, they pointed out, 'has of late years, at the cost of hundreds of thousands of pounds, been improved and modernized to such a degree as to bring it into a state of perfect efficiency' (*The Times*, quoted *C.T.J.*, 6 Sept. 1890, p.142). Outsiders retained their conviction that the Leblanc trade was obsolescent. Soon after the formation of the U.A.C., the *Chemical Trade Journal* (3 Jan. 1891, p.1) asked, 'Has the "fittest" triumphed over the unfittest and civilized him out of existence?' The calcium chloride wastes of the Solvay process, an observer wrote in *The Times*, had the advantage of being soluble and odourless, 'but waste of this kind is an offence to the soul of a chemist like Mr. Mond, even though his works pay in spite of it a dividend of 50 percent'. The attempt to make use of this calcium chloride was a triumph of applied science. 'There were plenty of other victories to be achieved, but they would fall to men similarly equipped [as Mond], and not to manufacturers content, as our manufacturers too often are, to proceed by obviously imperfect and wasteful methods without serious and adequate effort to reach higher things.' (*The Times* 21 Oct. 1892, p.8.) This was slighly unfair to U.A.C., but there was no denying that it operated an outdated process.

In 1890/1 U.A.C. made a net profit of £537,000 (in fourteen months' operation), and in the year 1892, £741,000. Soda ash prices (for 48 per cent soda ash) rose to £6 in late 1889 and stayed at that level till the end of 1892, but thereafter fell to £4. 15*s*. in 1894 and remained there for the rest of the century. These prices meant that the early levels of profits were never again approached during the nineties or indeed throughout the thirty-six years of operation of the U.A.C. with the single exception of the war-inflated profits of 1917/18 (Fig.25). In 1893/4 whereas Brunner Mond & Co. declared a dividend of 30 per cent on ordinary shares, U.A.C. managed only 5 per cent (*C.T.J.*, 3 Mar. 1894, p.145; 24 Mar. 1894). Their directors, Sir Charles Tennant remarked at the Annual General Meeting of March 1894, had 'piloted the ship through the waves with considerable success'. In May 1894 the various works were asked what savings they had made since 1890. At the next Annual General Meeting Tennant noted that the previous year had been an extremely bad one for the company; after forty years in the chemical trade he could not recall one in which the fall in prices had been so continuous.

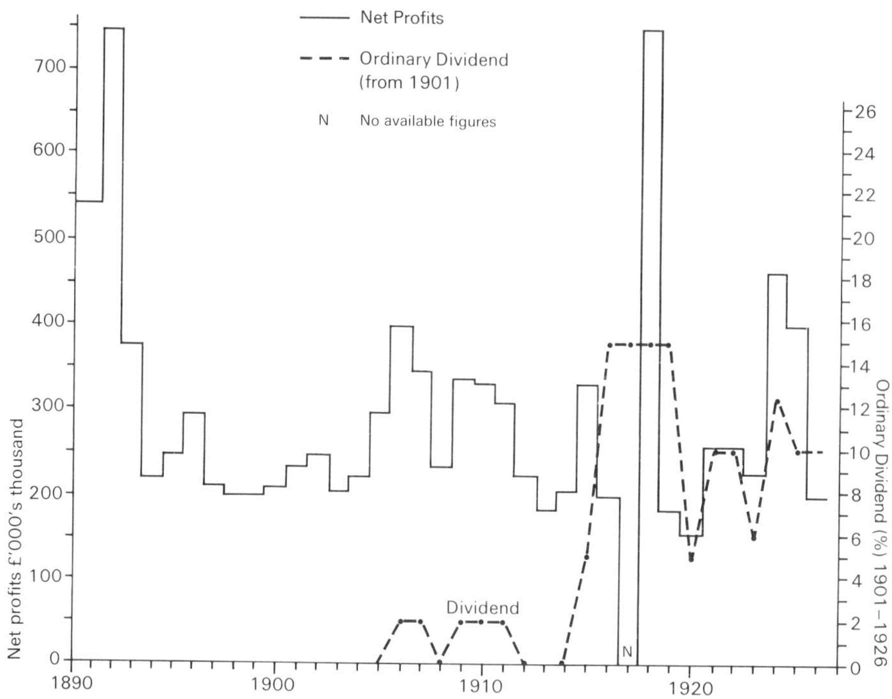

Fig. 25. U.A.C. profits and dividends 1890–1926 *Source:* I.C.I., Records

When Brunner Mond & Co. were mentioned during the meeting, a voice from the shareholders shouted, 'They can make their works pay when you cannot'. Another shareholder said that 'he had been connected with commercial matters since he was eight years of age, and everything he had touched had turned into money except what he had touched in that company, and he was sorry to say he had lost £1100 in twelve months'. (*C.T.J.*, 23 Mar. 1895, pp.192–5.) Things did not improve. At the Annual General Meeting in 1900 a shareholder argued that they ought to be earning 30 per cent like Brunners, 'There was too much patronage in the company, and it was overrun by officials and superfluous men' (laughter) (*C.T.J.*, 24 Mar. 1900, pp.204–5). In the summer of 1901 the journal *Engineering* contrasted Brunner Mond & Co. with U.A.C. The former were 'of course a standing monument to the success which may always be expected to attend business acumen, extensive scientific attainments, and last, but not least, the possession of the patent process which renders all opposition futile . . .'. It compared their dividends with 'the somewhat lugubrious balance sheets which have been issued year by year by the United Alkali Company since its inception . . .'. (*Eng.*, 12 July 1901, pp.54–5.)

Within nine months of its formation U.A.C. had begun to dismantle properties it considered antiquated, 'thus leaving room for the establishment of more suitable plant at a future date', as was optimistically stated. It also closed unremunerative sections of other plants and kept the rest on low production in order to bring supply into line with demand. (*C.T.J.*, 24 July 1891, p.41.) In March 1893, the chairman, Brock, reported that in the last year the process of concentrating production in those works and localities with lowest costs had continued (*C.T.J.*, 18 Mar. 1893, p.178). Trade conditions worsened, however, and profits declined further. By autumn 1894, throughout the United Kingdom the U.A.C. was laying off men and closing bleach and caustic soda departments for an indefinite period (*C.T.J.*, 17 Nov. 1894, p.308). By December 1894 sixty-seven production processes in all had been stopped in twenty-seven works, a major overhaul which included the complete closure of thirteen plants, almost one-quarter of the intial number. In 1891 U.A.C. directors considered the advisability of closing Muspratts' Liverpool works. A detailed report made in 1894 (Davies, 1894) showed that all its saltcake went abroad, much of its sulphuric acid to Widnes, and its hydrochloric acid to Ellesmere Port for Burnells galvanized iron plant. In 1896 the works was closed.

Table 55. *U.A.C. works acquired 1890/1 and closures 1890–1904*

	Acquired 1890/1	*Closures 1890–4*	*Closures 1895–9*	*Closures 1900–4*
St. Helens	12	2	2	—
Widnes	14	2	2	1
Runcorn & Weston	4	—	—	—
Flint	2	1	—	—
Liverpool	1	—	1	—
Tyneside/Durham	10	4	2	1
Glasgow/Irvine	4	1	—	—
Dublin	3	3	—	—
	50	13	7	2

Source: U.A.C., List of Works.

Of the original thirty works of the St. Helens, Widnes, Runcorn, and Weston group, only five closed in the first four years. In 1895 there were plans to shut down Wigg works in Runcorn, but it was re-equipped to produce cyanide and artificial manures. (*C.T.J.*, 31 Oct. 1895, pp.901–2; 18 Dec. 1897, p.391) Eleven St. Helens works came into U.A.C. Of these the St. Helens Chemical closed in 1891, Walker in 1892, McBrydes in 1895, and Sutton Lodge in 1896. In Widnes only Hay Gordon, Snape, and Liver had gone by 1896. The axe fell heavily on the Clyde group and even more heavily in the North-East.

In Scotland U.A.C. quickly closed Clydebank. At the North British Chemical Works, Dalmuir, dismantling was under way by early 1892 though much of the plant was said to be the best of its kind in the United Kingdom and was removed to other works. Make was severely restricted at Irvine and at

Eglinton, concentrating production at cheaper points—outside Scotland. (*J.S.C.I.*, 31 Mar. 1892, p.284; *C.T.J.*, 17 Nov. 1894, p.308.) Costs at these works were high (Fig.26). In the year ending 30 June 1896, saltcake costs at St. Rollox were the third highest out of twenty-three works making the product, and 62 per cent higher than the median. In other products it was a cheaper producer than its Scottish rivals, no doubt largely because of its greater scale (Tables 56 and 57). While run-down was going on in Scotland, Netham was reconstructed to increase its output. (*J.S.C.I.*, 31 Mar. 1892, p.284.)

U.A.C. acquired ten plants in the North-East, all except Seaham on the Tyne. Only four survived after ten years. Within one month of U.A.C.'s

Table 56. *Costs of sulphuric acid production at Scottish works 1893*

	Production (tons)	*Costs per ton* *s. d.*
St. Rollox	21358	12 4
Irvine	3737	15 5
Eglinton	1222	27 1

Source: U.A.C. Records

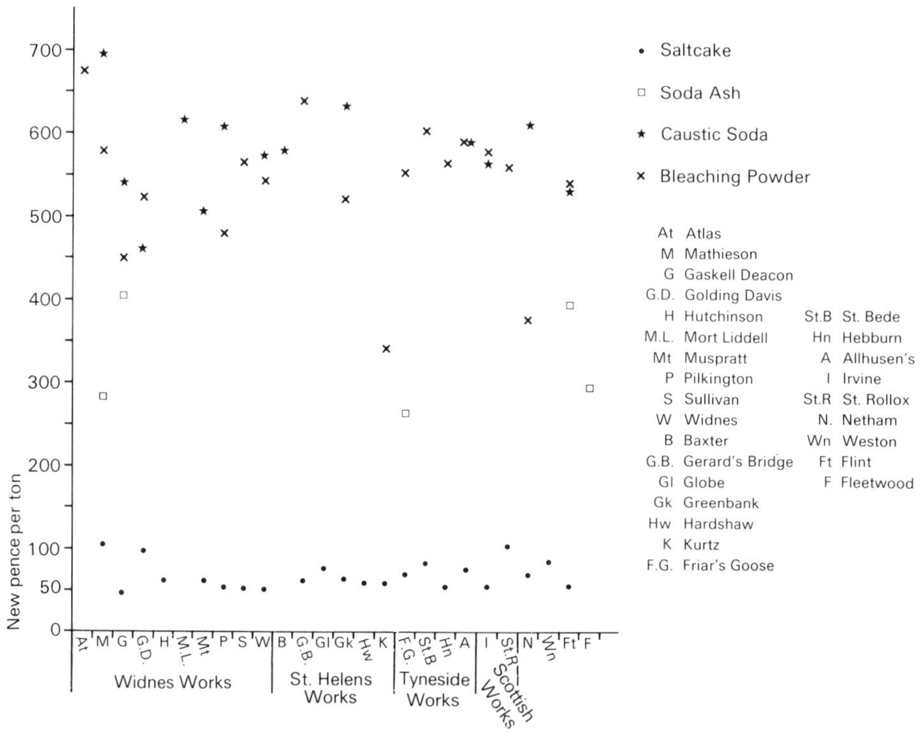

Fig. 26. Costs of production at U.A.C. works, 1896 *Source:* U.A.C., 1896

Table 57. *Costs, selling prices, and profits at Scottish works of U.A.C., 1896* (per ton)

	Costs		Selling price		Profits	
	s.	*d.*	*s.*	*d.*	*s.*	*d.*
Caustic soda						
Irvine	112	9	143	7	30	10
U.A.C. average	114	8.5	151	3	32	2.6
Bleaching powder						
Irvine	115	5	133	4	17	11
St. Rollox	112	0	131	11	19	11
U.A.C. average	104	8.5	131	0	26	4.3

Source: U.A.C., 1896.

formation dismantlers had been engaged to dispose of all the plant at the Tyneside Chemical Works in Felling, which had already been idle for five or six years. (*C.T.J.*, 6 Dec. 1890, pp.356, 358.) In the remaining works, costs were rather high and profit margins low. Thus in the first half of 1896 when the average profit of U.A.C. works on bleaching powder was 26*s*. 4.3*d*. a ton, it was 18*s*. 10*d*. at Friars Goose, 16*s*. 9*d*. at Hebburn, 11*s*. 3*d*. at St. Bede, and 10*s*. 11*d*. at Allhusens, the remaining works on the river. Allhusens' saltcake costs were about average, their caustic soda costs a little over, and their profits a little under average (U.A.C., 1896).

The North-West dominated U.A.C.: twelve of its first subscribers for shares were there, and only four on the Tyne (*C.T.J.*, 8 Nov. 1890, p.292); ten of the first twenty-three members of the U.A.C. Board were from Widnes. The strong links with the North-West, derived inevitably from its Liverpool headquarters, combined with cost and profit figures to encourage further run-down elsewhere and the concentration of expenditure in this district (Fig.27). At one of their first meetings, in January 1891, the directors of the Tyne Management Committee of U.A.C. stressed that retention of the soda crystal trade of London and of the Baltic was of the utmost importance to their works. Nevertheless, within seven years both U.A.C. and Brunner Mond & Co. had begun to make soda crystals in London, which by the early twentieth century consumed up to 80,000 tons of crystals annually. U.A.C.'s Clyde Wharf plant on the Thames incorporated equipment from Seaham Harbour and from St. Bede works. (U.A.C., Tyne, 20 Sept. 1897.) In 1895 the Liverpool board negotiated with Brunner Mond & Co. an equal division of soda crystal supplies to London and the South-West (U.A.C. Minutes, 6 Dec. 1895). In the first quarter of 1898 it was agreed that weekly crystal make should be restricted to 100 tons at Tennants, 400 at Friars Goose, 400 at Clyde Wharf. By the summer, however, the last had been extended to 500 tons after more equipment had been moved from the Tyne. (U.A.C., 28 Feb. 1898.) By 1899 exports of soda crystals from the Tyne were less than one-fifth of the 1890 level. U.A.C. sales declined in the London market even though the company had better water access to the Thames than Brunners. (U.A.C., Tyne, 10 Nov. 1902.) Meanwhile general decline on Tyneside continued. In

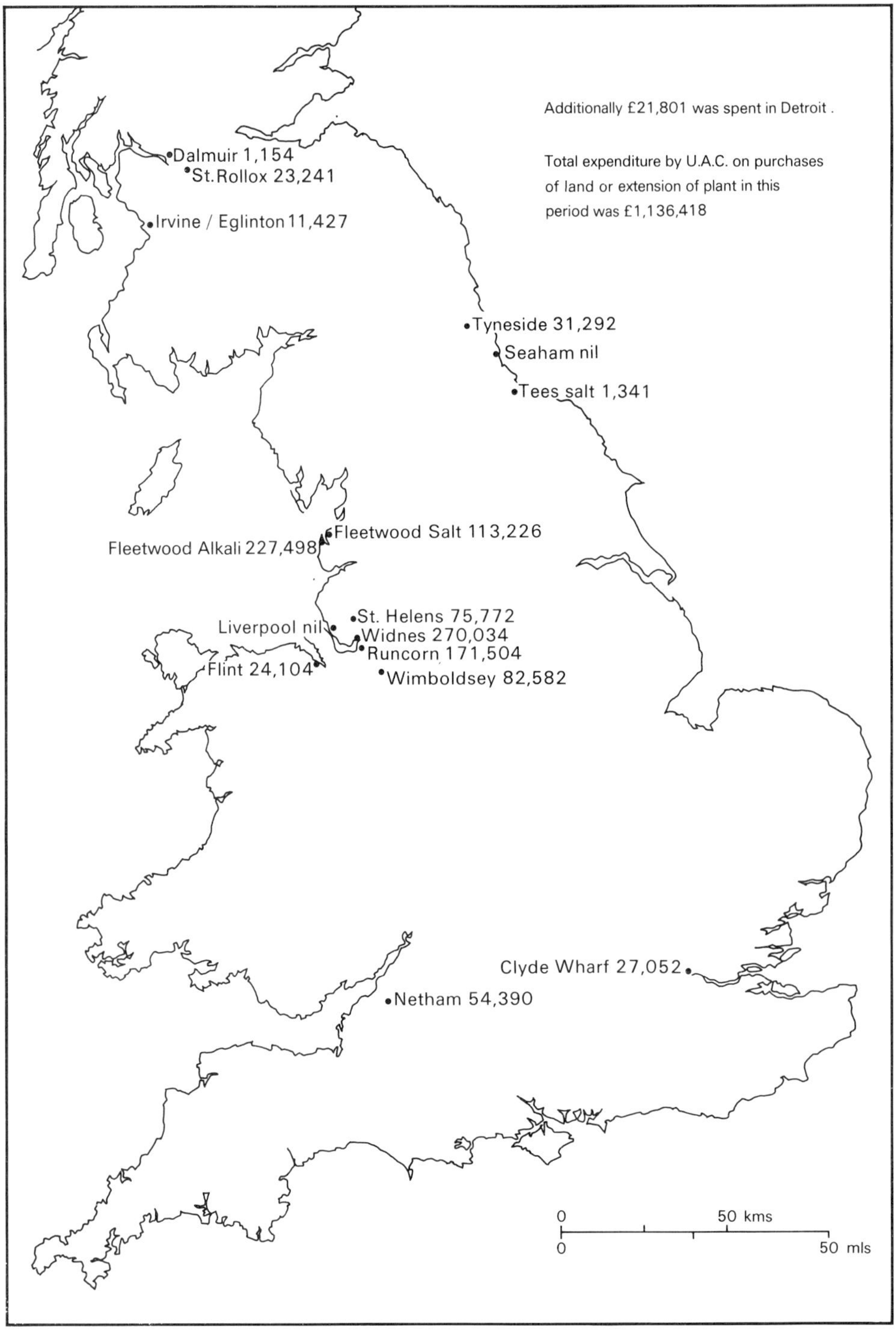

Fig. 27. Expenditure 1891–1911 on U.A.C. plants *Source:* U.A.C., Ledgers

1892 the General Management Committee decided that no more soda ash should be made there for sale. (U.A.C., Tyne, 1 Feb. 1891; 15 Dec. 1890; and 7 Mar. 1892.) In 1891 U.A.C. accepted an offer of £7,400 for Tyne Dock Works site (U.A.C. Minute Books). St. Bede stopped in November 1898 and by the following April had been almost completely dismantled (Table 58).

Table 58. *Weekly production or capacity for soda crystals at U.A.C. east coast works 1897/8 and 1911* (tons)

	1897/8 (production Dec. 1897–Feb. 1898)	*1911 (capacity March)*
Friars Goose	400	420
Tennant	100	150
Clyde Wharf	400	660

Sources: U.A.C. Tyne; U.A.C., 1911.

The creation of U.A.C. brought the expected advantages from linkage and specialization. Works that escaped closure were brought nearer to a common level of efficiency. By 1891 there had been eleven U.A.C. works in the St. Helens district, mostly concerned with alkali plants but including one copper works. 'At that time each works had its own method of working, from the "go as you please" style up to the latest methods.' They were all brought into line by the local managing director J. R. Wylde, who paid close attention to even the small details of cost economy. One outcome was that 'his district was specially noted for improvements and up-to-date methods'. (*C.T.J.*, 28 Jan. 1899, p.67.) Even so, not all were profitable (Table 59). Plant linkages were important too. In the spring of 1891 it was agreed that Richardsons' works at Jarrow should stop decomposing salt and obtain its saltcake instead from other works. In 1893 Tyneside works were using some Scottish saltcake, and a little later for soda crystal manufacture at Friars Goose and Hebburn they were working on Fleetwood ammonia soda ash. (There were complaints on the Tyne that the works could not make a profit at the price at which it was supplied—£2. 19*s*. 5*d*. at Fleetwood but £3. 8*s*. 0*d*. delivered on the Tyne (U.A.C., Tyne, 14 Dec. 1896).) By 1901, when saltcake was being transferred from one works to another on the Tyne for 30*s*. a ton delivered, it was decided that saltcake should be delivered from Glasgow to the Tyne at no higher price than this (U.A.C. Tyne).

Table 59. *Anticipated profits and losses of St. Helens district works, 1900* (£)

	Profit	*Loss*
Baxters	26,000	
Globe	500	
Gerards Bridge		3750 (on saltcake)
Hardshaw Brook		3750
Kurtz	750	
Greenbank	12,000	
McKechnie	30,000 (copper)	
	£69,250	£7500

Source: U.A.C. 1903 (I.C.I., London Records).

Matching the movement of material was the concentration of production (Fig.28). Between 1896 and 1911, from being spread over thirteen U.A.C. works, caustic soda production was concentrated in eight, one of which—the Kurtz works in St. Helens—seems to have taken up the trade for the first time. In the same period bleaching powder production was concentrated in thirteen works instead of sixteen, three being new to the trade. (U.A.C. 1896 and 1911.) On the Tyne, Allhusens concentrated on caustic soda (in which indeed it was by far the largest works in U.A.C.). It was also by 1911 the group's biggest producer of bleach. Its surviving local rivals, Friars Goose and Tennants of Hebburn, were much more prominent in soda crystals, and the latter was also important in bleaching powder (Table 60).

Table 60. *Bleaching powder production on the Tyne, 1894–1911* (tons per week)

	1894 (March)	*1899* (November)	*1901* (March)	*1911* (March capacity)
Allhusen	309	320	125	326
Tennant	163	200	110	140
Friars Goose	144	130	70	150
St. Bede	70	(closed 1898)	—	—

Sources: U.A.C. Tyne, various, and U.A.C. 1911.

As plant specialization and closure continued, equipment from old works was moved widely. In 1891, so that the Richardson works could be closed, the hyposulphite plant was moved to Allhusens. A year later soda crystal plant was moved from Jarrow and from Heworth to Hebburn and to Friars Goose, where capacity was being extended. In 1893 the remaining plant from the Jarrow Chemicals' Jarrow works was moved to Allhusens, St. Bede, and Hebburn. Five years later, as St. Bede closed, furnaces were moved to Friars Goose, which survived for another seventeen years. Plant was moved both within the various districts and more widely. Less than a year after its formation, U.A.C. had already moved Tyneside plant to Irvine works and to the three St. Helens works of Greenbank, Sutton Lodge, and Globe. (U.A.C., Tyne.) In 1896, when it was decided to build a 125 ton per week soda crystal plant at Fleetwood, the Liverpool directors advised 'utilizing as far as economical part of the plant at Gaskell Deacon (Widnes) or other works' (U.A.C., Liverpool).

It was not surprising that a nationwide organization should lose the intimate contact with neighbourhoods that firms with individual works had had. Throughout its existence, as U.A.C. faced difficulties and had to make decisions about closure or running-down, it often seemed to deal harshly with its employees and with the localities in which its works were located. In Widnes, which remained its chief centre of operation, the number of workers employed fell from 5,760 in 1894 to 4,221 in 1900 and to about 3,600 on the eve of World War I (Hardie, 1950, pp.150–1). In October 1894 the Tyne directors resolved to cut employment in the mechanical and traffic depart-

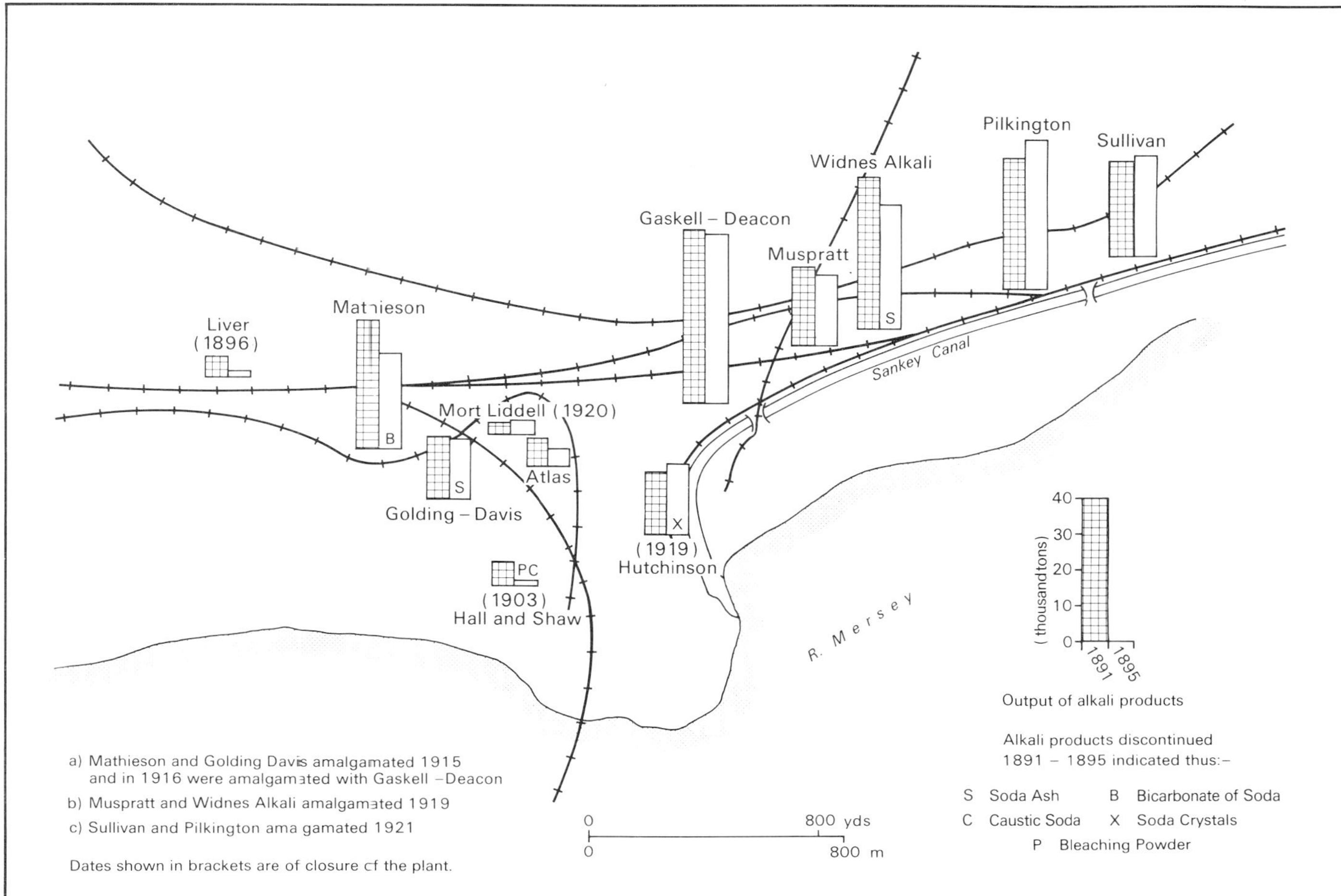

Fig. 28. Rationalization in U.A.C. Widnes Works 1891–5 *Source:* U.A.C., 1896b

ments of Allhusens and 'by continual application, prudently and carefully directed' to bring about a gradual 'reduction in the amount of wages paid weekly'. Two years later they were still trying to cut wages, while demand for labour on the river generally was very active and advances were being made. In these circumstances, the men refused both a lengthening of hours of work and a reduction in wages. Eventually it was agreed to compromise. (U.A.C., Tyne.) The company could and did act ungenerously when there was no danger of losing labour as a result. For instance, in 1896 a public meeting was held in Newcastle to aid the proposed new infirmary; £38,504 was raised. 'Amongst the subscriptions are some £10,000 contributed by the firm and partners of Sir W. G. Armstrong and Co. Limited and £100 by the U.A.C. Limited, who have extensive works in the neighbourhood' (*C.T.J.*, 17 Oct. 1896, p.245). The Minutes of the Tyne directors for 9 November 1908 contained the following: '*Distress at Hebburn*. The Hebburn Schools' Canteen Committee have made repeated and urgent application for assistance towards a fund for providing meals for necessitous children attending schools in the district. The Tyne office with the consent of the G.M.C. propose to contribute the sum of £5. . . .'

The Tyne provides another illustration of the old and the new approaches to localized distress. James C. Stevenson had joined Jarrow Chemical Works in 1843. In 1867 he had been mayor of South Shields, and was Liberal M.P. for the town from 1868 to 1895. Stevenson was a member of the committee which established U.A.C., and, until May 1891, chairman of the Tyne District Directors. A month later, after his chairmanship had ended, the Tyne directors decided that in accordance with the wishes of the Liverpool board, Jarrow works should be closed at the earliest possible moment. T. W. Stuart was instructed 'to proceed accordingly, dismissing the workmen as may be necessary and referring the office staff to Newcastle' (the class distinction will be noted). At the beginning of July Stevenson telegraphed from Chamonix requesting a three-month delay in closing the Jarrow works. The Committee refused (Fig.29).

The Fleetwood ammonia soda works

In the mid-1890s, U.A.C. brought an ammonia soda plant into operation at Fleetwood. This development was the outcome of two factors: the acquisition of salt and the technology that U.A.C. inherited. In the 1870s during boring for hematite a bed of rock-salt 300–500 foot thick was discovered at Preesall. In 1889 this field was explored and developed by the Fleetwood Salt Company which in the following year was absorbed by U.A.C. Two years later it was decided to build alkali works at Burn Naze. The technology for the new plant was derived from one of the constituent companies of U.A.C. in Widnes.

In the late 1880s Neil Mathieson built an ammonia soda plant on the Widnes Marsh to work a process patented by the Pole, Hawliczek. The process was installed at Fleetwood by the U.A.C. It proved less satisfactory than

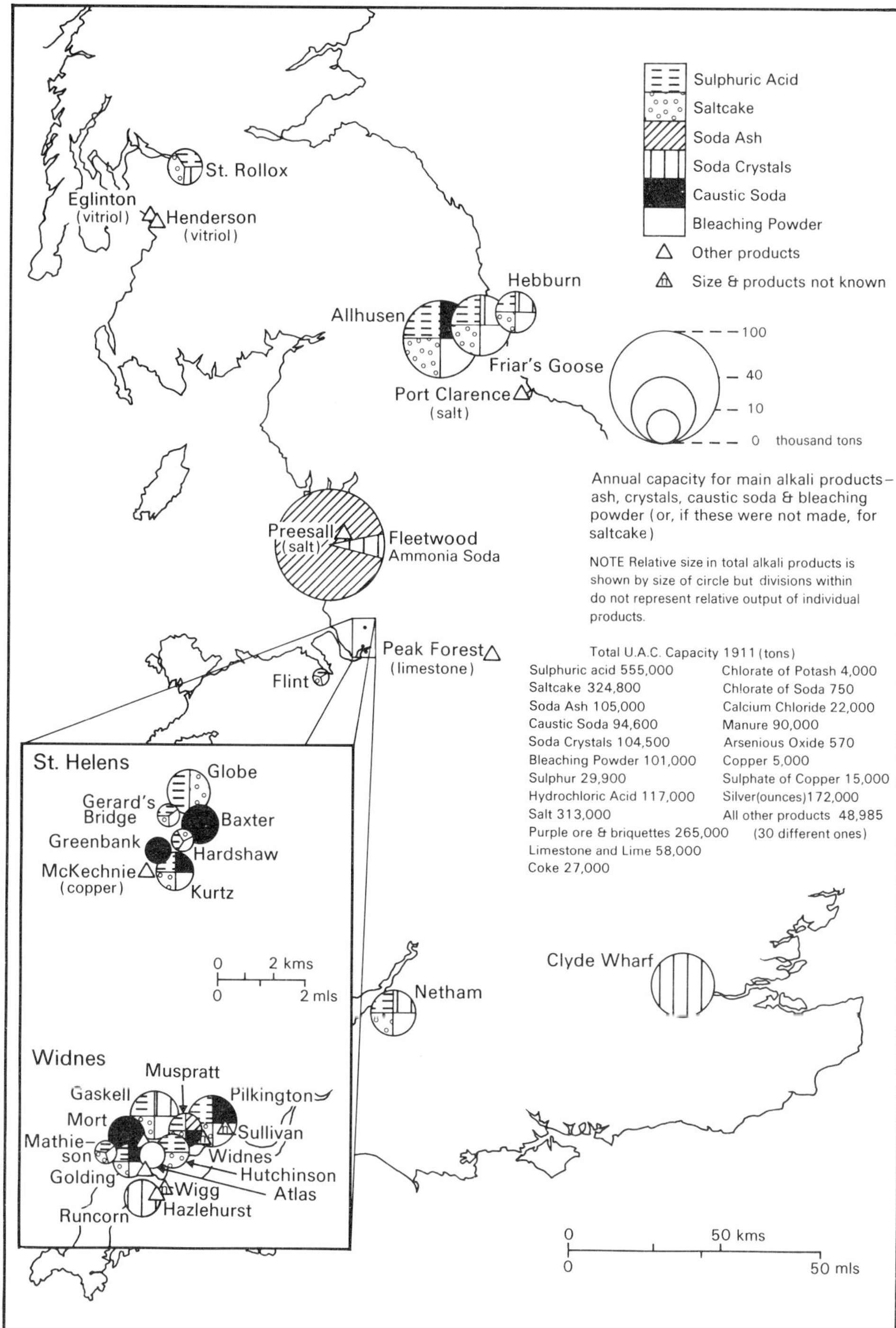

Fig. 29. U.A.C. plant capacity 1911 *Source:* U.A.C., 1911

Solvay's. Other factors also inhibited Fleetwood competition with Brunner Mond. The brine was weaker than that in Cheshire and therefore required more fuel in processing, and the salt bed was deeper, though not nearly so deep as on Teesside. There were considerable difficulties in the early years of operation, and in the first half of 1896 material costs per ton of soda ash, though well below Mathiesons', were cancelled out by higher process costs than at the Marsh works. (U.A.C., 1896; Fleetwood, 1903; Fleetwood, 1963; Table 61.)

Table 61. *Material costs and total cost at Fleetwood and at Mathiesons, 1896*

	Fleetwood			*Mathieson*		
	£	*s.*	*d.*	£	*s.*	*d.*
Raw materials	1	10	7	1	13	1
Total costs	2	18	7	2	15	4

Source: U.A.C., 1896.

For these and other reasons connected with the differences in the nature of the two companies, Fleetwood output grew much less rapidly than Brunner Mond's capacity. In 1909 Brunners made 291,000 tons of soda ash and in 1914 414,000 tons; Fleetwood's 1911 capacity was 100,000 tons. When in 1926 Fleetwood was brought into I.C.I., the new owners rebuilt it to operate Solvay's process. By the late 1930s the five ammonia soda works brought in from Brunner Mond had a daily capacity of 3,350 tons, Fleetwood of only 400 tons. (U.A.C., 1911; Reader 1970, p.217; Hou 1942, p.43.) Fleetwood closed in 1964, and thereafter Winnington, Wallescote, and Lostock, all within three miles of the centre of Northwich, remained as the only plants producing soda ash. This perhaps unique degree of concentration of a basic industry in Britain is the more remarkable in that this is where the industry began a century ago.

CHAPTER 14

United Alkali Company: rivals and customers

The United Alkali Company was in part formed in order to try to check the headlong decline of alkali prices. In the fifteen years to 1890 soda ash prices had declined by 50 per cent. (Dick, 1973, p.123.) The amalgamation had some success in achieving its aim; prices continued to fall in the first ten years but at a very much slower rate. After 1900 there was a slight increase and then a more rapid fall. It is, however, important to remember that the Leblanc and U.A.C. share of British production was continuing to fall rapidly (Fig.30). (In 1890 both Brunner Mond & Co. and U.A.C. output was 180,000 tons of soda ash. By 1907 U.A.C. produced 60,000 tons; two years later Brunners' output was 290,000 tons. In 1911 U.A.C. capacity was still 105,000 tons, but by 1914

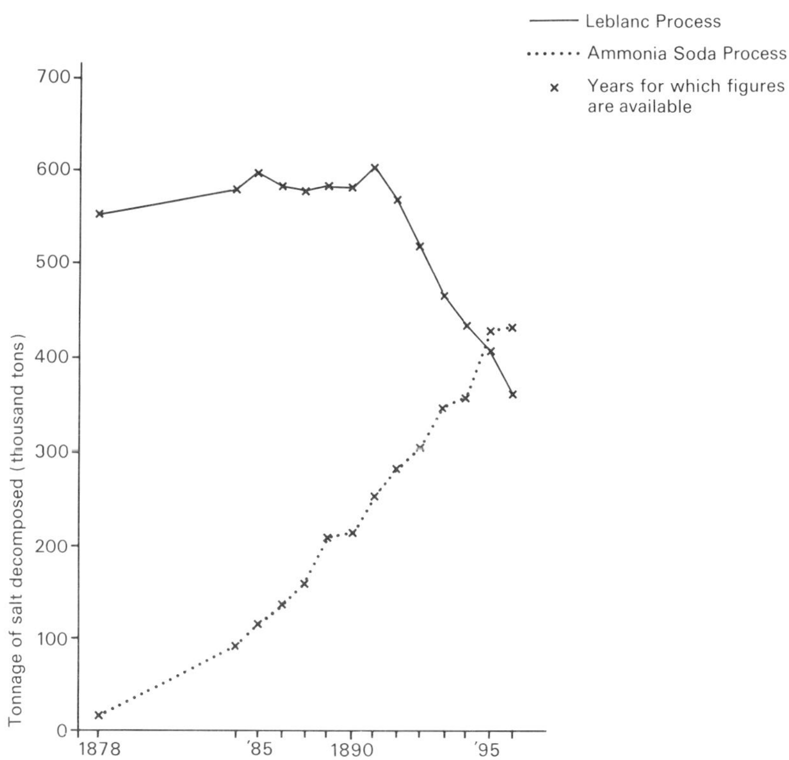

Fig. 30. Salt decomposed by Leblanc and ammonia soda works 1878–96 *Source:* Alkali Inspectors' Returns

Brunners produced 414,000 tons.) (U.A.C., various; Haber, 1958; Reader, 1970, p.217.) When agreement kept soda ash prices for ten years at £4. 15*s*. a ton, Brunners were profitable at that level and U.A.C. was not (Haber, 1958, pp.183–4).

In bleaching powder also, on which U.A.C. depended more heavily, prices were uncertain but in general drifted definitely downwards. The U.A.C. attempt to maintain prices by agreements with its major rivals and by the elimination of smaller producers met with many difficulties.

U.A.C. relations with rivals

U.A.C. had to be willing to treat with its major rivals. Early in 1892, for instance, it made an agreement with Chance and Hunt, by then the biggest independent Leblanc firm (Chance & Hunt, 1892). Its most important relationships, however, were with the ammonia soda firms and therefore particularly with Brunner Mond & Co. and Solvay. Agreement was difficult to reach, for these firms commanded the cheapest process, had a 'scornful and suspicious attitude to U.A.C.', and 'discounted U.A.C. as serious competitors for the prosperity of their alkali business except for occasional annoying excursions by U.A.C. in their important alkali export trade' (Dick, 1973, p.30). Brunners would not negotiate with Leblanc firms in the eighties, but U.A.C.'s size forced them to do so later. In 1891, in exchange for U.A.C. agreement not to make more than 15,000 tons of ammonia soda, Brunners agreed not to make more than 165,000 tons of soda ash (Reader, 1970, p.109). The obvious move for U.A.C. was to edge out of soda ash and move further into caustic soda and bleaching powder. Unfortunately, from 1895 onwards, this brought them into direct confrontation with electro-chemical processes.

In spring 1895, the U.A.C. board decided they were willing to cut soda ash production and to extend bleaching powder capacity on the Tyne and in Lancashire, provided Brunners would not increase chlorine production over the previous year's level and would not produce caustic. They also wished Brunners to relinquish their idea of building a Solvay works on the Tees. In return U.A.C. would join Brunner Mond & Co. in subsidizing the withdrawal of Bell Brothers and Bowman Thompson from the trade and the closure of their works. (U.A.C., Liverpool.) Negotiations were broken off, resumed, and finally carried through later in the year. There is, however, evidence of a continuing jockeying for position. In 1894 Brunners had opened the Silvertown soda crystal plant to process ash shipped from Winnington. In 1897 U.A.C. brought into production their soda crystals plant at Clyde Wharf, also in Silvertown. The agreement of autumn 1895 involved a U.A.C. commitment to sell the Bowman Thompson chemical production. Instead of removing their output, U.A.C. caused large tonnages to be stocked at Lostock Gralam, and in August 1896 had to be called upon to remove them (*C.T.J.*, 22 Aug. 1896, p.114).

In January 1896 the U.A.C. board decided to extend Fleetwood to a capacity of 1,200 tons of soda ash weekly or some 60,000 tons annually. The next year discussions took place with Solvay over soda ash. Meanwhile, as tariffs and Solvay capacity both went up overseas so even Brunners' exports of ash suffered, their loss of business being especially serious in the United States. This was the background to discussions that took place in July 1900 at Brunner Mond & Co.'s London office. Solvay representatives were also there.

U.A.C. were led by John Brock, Brunners by Sir John Brunner and Alfred Mond. The account which follows is based on a United Alkali report. Opening the discussions, M. Armand Solvay suggested that it was time that they renewed their friendly agreements. He suggested, in the words of the U.A.C. report:

> Now that Bowman Thompson and Company had disappeared [taken over in 1900 by Brunners] we ought to try to make a more definite agreement—hitherto we had enjoyed what could only be called an armed peace, which was not good for us inasmuch as it compelled him (that is each of them) to arm himself with modern arms and plenty of ammunition and that we could have had much better terms before than we could expect now if we would have agreed before to limit our production of ammonia soda, and that the only way to make a really lasting and good agreement was for the U.A.C. to agree to limit its production of ammonia soda and for a long period.

The fears of Solvay and of Brunners had been increased by the fact that in January 1900 U.A.C. had spent £80,000 in buying Wimboldsey Estate one mile south-east of Winsford, so securing one of the last major salt-fields in Cheshire. This was a rational move, for U.A.C. were then buying some 120,000 tons of salt from independent concerns. (*C.T.J.*, 24 Mar. 1900, pp.204–5.) With the London and North Western line, the Shropshire Union Canal, and the River Weaver running through it, Wimboldsey seemed excellently sited for the construction of an ammonia soda works, and contemporaries expected U.A.C. to build there. (U.A.C., Liverpool; *C.T.J.*, 27 Jan. 1900, p.79.) In the July discussions, John Brock replied to Armand Solvay's remarks by the non-committal statement that U.A.C. were willing to agree on equitable terms. Alfred Mond reacted with a fierce assertion of his firm's position and a fighting reference to the Wimboldsey purchase. Brunners maintained that 'the Ammonia Soda process must be considered to be theirs', and no agreement was possible unless U.A.C. agreed to limit production. He noted that 'we had lately purchased a large estate in Cheshire and they could not believe that we bought it simply to make salt for the Leblanc process!', and added that 'if ever the U.A.C. proceeded to the erection of an ammonia soda works in Cheshire it would be the duty of Brunner Mond and Solvay to their shareholders to do all they could to prevent those works ever coming into active operation'. In response to this outburst, Brock explained that at the moment there was no thought of either building ammonia soda capacity at Wimboldsey or of extending it at Fleetwood. Sir John Brunner then asked if U.A.C. would suggest a figure for limitation of make at Fleetwood and U.A.C. replied that if any more Leblanc soda ash capacity had to be closed down they would have to extend there in order to keep up their make of caustic soda. Mond stressed that loss of United States business had rendered much of their soda ash capacity idle and it was ridiculous for them to have to build up caustic capacity to employ more of their existing capability to produce ash, while U.A.C. extended ammonia soda capacity to support existing caustic plants. He suggested Brunners supply U.A.C. with soda ash at a reasonable price if U.A.C. would not extend ammonia soda capacity. 'This', as U.A.C. recorded it, 'was their great point, put a limit to your production and we will make a bargain.' The U.A.C. representatives agreed to recommend to their board that Fleetwood capacity be limited for five years to 80,000 tons of soda ash and 5,000 tons of bicarbonate of soda. In return, the bleaching powder and caustic soda capacity of Brunner Mond & Co. should not be increased. Their representatives declared the 80,000 level for Fleetwood ridiculous since in 1899 its output had been no more than 46,000 tons. Brock defended the figure, explaining that currently their make was equal to 62,000 tons a year. Brunners seemed willing to accept a 62,000 ton ceiling, but at this point the meeting ended. Ten days later an agreement was concluded between U.A.C., Brunner Mond & Co., and Solvay and Company (on behalf of the Deutsch Solvay Werke), whereby Brunners agreed to limit caustic and bleaching powder production for five years and U.A.C. committed themselves to limit ammonia soda production to 70,000 tons and to refrain from building another works (U.A.C., 1900A). In fact, up to 1901, Fleetwood was still having what were called 'teething troubles' and it was reckoned that 1903 costs could be £12,000 below the 1902 level. (Fleetwood, 1903; Fleetwood, 1963.) After the expiration of the agreement, Fleetwood was extended, so that by 1911 it could produce 7,500 tons of bicarbonate of soda (the whole of U.A.C.'s capacity for this product) and 100,000 out of their total of 105,000 tons of soda ash. (U.A.C., 1911.)

The agreements limited but did not end trouble between U.A.C. and their rivals. The run-down of the company's plant went on (Table 62.)

Table 62. *U.A.C.: number of works making various products 1896, 1911, 1925*

	1896	*1911*	*1925*
All chemical products	38	31	18
Soda ash	5	2	1
Saltcake	23	18	3
Caustic soda	13	8	4
Bleaching powder	18	14	?

Sources: U.A.C., List of Works; U.A.C., 1923/26 (Comparisons of Works).

An example of a late agreement entered into by U.A.C. was that of October 1919 between Brunner Mond & Co., the Castner-Kellner Alkali Company, and U.A.C. (1919). It applied to the whole world except the United States and Canada, the continent of Europe, and the rest of the Mediterranean, Black

Sea, Levant, Malta, and Gibraltar. For products made by U.A.C. and Brunner Mond & Co. it was agreed that the proportions of total deliveries to the areas within the terms of the agreement should be as listed in Table 63, vividly illustrating the eclipse of what had initially been the world's leading chemical firm.

Table 63. *Apportionment of deliveries by Brunner Mond & Co. and U.A.C., 1919* (percentages)

	Brunner Mond & Co.	*U.A.C.*
Soda ash	82	18
Bicarbonate of soda	85	15
Soda crystals	55	45

Source: Brunner Mond, 1919.

As regards the British market, there was an interesting section dealing with deliveries of hydrochloric acid. Brunners agreed to give up its manufacture. In consideration of this U.A.C. was not to deliver the acid in a so-called Midlands District within 30 miles of Birmingham Town Hall. The tonnage being delivered there by both U.A.C. and Brunner Mond & Co. was to be transferred to Chance and Hunt, Brunners in turn undertaking to arrange that Chance and Hunt should not deliver to the north of a line drawn through Northwich, Macclesfield, and Buxton, or try to extend their present territory northwards.

U.A.C. and innovation

One of the most interesting aspects of U.A.C. was its attitude towards research and development work in heavy chemicals. When formed it employed 12,000 but had only 50 chemists (Haber, 1958, p.156). The Central Research Laboratory which it established at Widnes in 1891 was unusual in any chemical company at that time outside Germany. In charge of it was Ferdinand Hurter, a chemist of very high ability. Born in Schaffhausen, Hurter came to England in 1867 to work for Gaskell-Deacon and Company. In 1890 he became chief chemist and scientific adviser to U.A.C., a post which he held until his death. Hurter was devoted to his work, but was hampered by two factors: an apparent in-built antagonism to his more successful German rival Ludwig Mond, and perhaps too great a deference to U.A.C.'s rather timid management. He was a committed defender of the Leblanc process: 'Hurter's able mind was not alone taken up in the matter, his emotions too were engaged. As others might defend a political creed or a philosophic view, Hurter defended his process; in that defence he found his vocation.' (Hardie, 1950, p.164.)

In his chairman's address to the U.A.C. Annual Meeting of 1898, John Brock spoke of great changes in the industry: 'In fact, since this company was formed, a great revolution has passed over the alkali trade, and few but those who sit at the directors' table can properly appreciate the changes which have taken place' (*C.T.J.*, 26 Mar. 1898, pp.220–1). There was little evidence then or later that the U.A.C. board was ready to respond to the challenge. Indeed, Brock's speech carried a strong flavour of despair, of an old industry with little eagerness for modern technology. After an unpromising start, he went on to speak of new technologies and to reveal the U.A.C.'s cautious attitude towards them:

> Almost every day sees a new process in the chemical trade, and especially is this so of electrical processes, which many people seem to be prepared to believe to any extent. All such processes, however, require the most careful expert examination, because few of them are of value as a means of producing cheaply; in fact, scarcely any of them at present working have been run on a manufacturing scale a sufficient time for it to be known what is the real wear and tear, and therefore the real cost of manufacturing by them.

A year later Brock was again arguing that their problems were due to unreasonable tariffs and a lack of 'fair play' rather than to obsolete processes and plant. In the past year they had extended in fertilizers and in cyanides—one of the undoubted technical successes of the Central Laboratory—but he was scathing about new processes:

> We still have the 'ubiquitous' inventor to deal with, and one is almost disposed to vote him a nuisance; but he has his use and merits, and must have our careful attention. But it is really surprising how little sound grain there is in what is offered to us; still some of these grains may turn out golden, though they require a good deal of careful winnowing. (Report of Eighth Annual General Meeting, *C.T.J.*, 25 Mar. 1899, pp.204, 205.)

Brock returned later to a lack of 'fair play' and the closure of markets, though by 1902 processes of operation were being recognized as deficiencies. What U.A.C. wanted, he said, 'was a wider market for their goods. They had plant capable of producing much more than could at present be sold, and they were leaving no stone unturned to open up new markets, and to develop new processes and products.' (Report of the Eleventh Annual General Meeting, *C.T.J.*, 29 Mar. 1902, p.289.) In no line were Hurter's technical short-sightedness and the conservatism of his Board clearer or more unfortunate than in relation to electrolytic processes of manufacture.

The Chemische Fabrik Elektron A.G. plant at Griesheim was at work in 1889/90. Hurter was soon involved in a study of the process, but after a quarter of a century in Leblanc alkali, he seems not to have been wholly open-minded. He was subsequently convinced by the mounting evidence, but a little later, and at a critical point, reverted to his earlier views, and so played an important role in the failure of U.A.C. to participate in the adoption of the new process. Another major factor, however, was the reluctance of those more closely associated with electrolytic alkali to let go their interest to the larger group. In April and September 1891 Hurter wrote reports on two early electrolytic processes, those of Greenwood and Le Sueur. In April he concluded, after considering operating costs, 'As an operation for the manufacture and sale of solid caustic soda and bleaching powder I still think electrolysis hopeless'. In September he decided that the over-all cost of Le Sueur caustic soda would be £4. 10*s*. a ton, or over £3 a ton more than with established U.A.C. processes. However, by November 1893 Hurter had submitted another report, asking for £7,000 to install dynamos and cells for large-scale experimental work. (Davidson, 1932.) In the autumn and winter of 1893/4 came the decisive period when the Castner process was considered by U.A.C. and allowed to slip out of their grasp.

On 24 November 1893 Hurter wrote to Brock concerning Castner's process: 'Castner's process is exceedingly ingenious and (I fear!) offers only practical difficulties.' He ended his letter: 'If you can obtain permission to see the process, I think it is our duty to see it and to try it.' On 1 December G. W. Balfour of the Aluminium Company of Oldbury, which held the Castner patents, wrote to Sir Charles Tennant, U.A.C.'s President, that they had no objection to Brock and Hurter visiting Oldbury to see the process. By the middle of the month U.A.C. had suggested that they should build an experimental plant for the process. The Aluminium Company decided instead to build the plant at Oldbury. Balfour however sent another letter saying, 'At the same time you may rest assured that we shall not forget the fact that your company was the first to inspect the process, nor the courtesy with which you have treated us.' In mid-January Brock observed to Hurter that Castner reckoned his process could produce caustic soda for £13 a ton whereas the U.A.C. costs were £13. 9*s*. 8*d*. U.A.C. was at this time ready to purchase Castner's patent and proceeded to draw up heads of agreement. At this critical point, however, Hurter wrote to Brock in a tone highly critical of the proposals. The demands made on U.A.C. were, he said, unreasonably heavy; the proposed royalty was too high; and it ought to be shown that the costs of Castner's process were much below those of U.A.C.'s processes and other electrolytic methods in order to make it worth while. He did not like expenditure on experimental plant at Oldbury where U.A.C. was unable to make a close assessment of fuel and labour costs. This was followed in February by further proposals from the Aluminium Company and counter-proposals from U.A.C. In June Balfour was writing of the possible reopening of negotiations but still insisting that experiments should be at Oldbury not at a U.A.C. works. In mid-July he wrote to say that the experimental plant had been built at Oldbury. (Electrolysis, 1893/4.)

The competition of electrolytic alkali soon began to be felt—as early as 2 October 1894 a sample of caustic made by Castner's process was laid on the table at the weekly meeting of the Tyne directors who, in Allhusens' works, controlled U.A.C.'s biggest caustic operation. In October 1895 the Castner-Kellner Alkali Company issued its prospectus. When the company's new works at Weston came into operation in 1897, it was producing 21,000 tons or just about one-ninth of the U.A.C. level of production in 1896. By 1911 U.A.C. capacity was 95,000 tons of caustic soda and 101,000 tons of bleaching powder; three years later combined Castner-Kellner capacity was still only some 52,000 tons of these products, but, in contrast to U.A.C.'s poor commercial record, the electrolytic company had been eminently successful. (U.A.C., 1896; U.A.C., 1911; Brunner Mond, 1926.)

In spite of the crisis over the formation of Castner-Kellner, U.A.C. remained complacent or opposed to innovation. In March 1894, just after the period of most vigorous negotiations with the Aluminium Company, Sir Charles Tennant's reply to a shareholder's question concerning electrolytic competition exemplified their attitude. While the new processes might come in some day, he did not think shareholders should make themselves unhappy about the Electro-Chemical Company (of St. Helens): 'Sufficient for the day was the evil thereof' (*C.T.J.*, 24 Mar. 1894). Such was the Board's ignorance of technology that, almost four years later, Hurter had to prepare a document for them explaining in simple terms the nature of electrolysis, with examples and answers (Hurter, 1898). Eventually both Allhusens and the Pilkington works at Widnes used electrolytic processes to make caustic soda (British Association, 1923), but the opportunity for U.A.C. to become the dominant producer of electrolytic alkali had passed.

U.A.C.: smaller rivals and customers

It was essential for a great commercial combine to try not only to eliminate rivals that were too small to command accommodation but also to retain and to extend its own customers. Neither was easy for U.A.C. while the maintenance of prices remained of prime importance. Steady, remunerative prices encouraged newcomers and the very existence of such a large firm as U.A.C. suggested that plant promotion might later be turned to advantage in a sell-out. The history of U.A.C. began with blatant attempts to eliminate small independent concerns. In December 1890 at the meeting of the directors it was noted that the Langdale Manure Company on Tyneside had sold saltcake to the Jarrow Chemical Company; the decision was then taken to telegraph Mr. Richardson in Newcastle and request him to undersell the Langdale Manure Company by 2*s*. 6*d*. a ton (U.A.C., Liverpool, 19 Dec. 1890, p.17). In 1883 the alkali trade of Wallsend had been revived when John Allen's old works was acquired by a new group, renamed the Wallsend Chemical Company, and then extended. One of its new lines was soap. (Campbell, 1962, p.37.) Immediately after its incorporation into U.A.C. it was decided to negotiate its closure with Thomas Hedley. In late December U.A.C. wrote to Hedley agreeing to stop and dismantle the soap-works in consideration of a cash payment of £5,000, with a view to selling the plant to the highest bidder 'stipulating that it shall not be re-erected on Tyneside unless its purchase was effected by your goodselves'. Negotiations seem to have gone badly, for at the end of February the plant was leased to Johnson Dodds and Company (U.A.C., Tyne). In autumn 1891 the Northumbrian Chemical Company was registered with a capital of £60,000 and with the former secretary of the St. Bede Chemical Company (now controlled by U.A.C.) as its managing director. From the start it seems to have been little more than a promotional exercise. (*C.T.J.*, 28 Nov. 1891, p.339; 26 Dec. 1891, p.404.) U.A.C. rose to the challenge. At their meeting on 10 May 1895, the directors in Liverpool authorized the Tyne Committee to come to an agreement with the Northumbrian Chemical Company whereby the latter's works should be 'entirely closed and made idle' for two years subject to an annual subsidy from U.A.C. of £7,500, the Northumbrian sales of bleaching powder passing to U.A.C. Almost two years later authorization was given for the completion of arrangements, U.A.C. now paying £25,000 in five annual instalments, and supervising the breaking-up of the works; the land was not to be used again for a chemical works. In March 1897 it was agreed that the sum to be paid should be £40,000. (U.A.C., Liverpool.) The Minutes of the Tyne directors revealed that the agreement was that for twenty years from 1897 the Heworth Shore land should not be used again to make any products turned out by U.A.C.—in fact, like much of the chemical shore of Tyneside, this area remained derelict for much longer.

With their customers or major suppliers, U.A.C. were in a weaker position. In 1903/4 they acquired copper pyrites deposits in Spain in order to safeguard their increasingly important sulphuric acid capacity. In the mid-nineties they had difficulties with the Salt Union, still a major source of supply in spite of the opening of the Fleetwood salt-field and the important expansion of Teesside salt production. The ammonia soda process had increased the decomposition of brine between 1880 and 1896 from 27,000 to 429,000 tons; but this led to a decline in the salt consumption from the Leblanc process. Demand for Salt Union salt for chemical purposes fell in 1890–5 from 600,000 to only 275,000 tons. Reacting to this in 1895 and 1896, just as U.A.C. entered strongly into competition in salt sales, the Salt Union threatened to begin alkali production themselves; they would operate the newest plant, and, their chairman added, be 'free from the burden of old and obsolete chemical works'. (*C.T.J.*, 7 Mar. 1895, p.162; 29 Feb. 1896, p.148.) Although this threat never materialized, U.A.C. were less successful with their customers.

Soap-makers began to make their own caustic soda at the end of the nineteenth century—as they needed it liquid they could save the considerable expense of evaporating it if they produced it at their works. By 1901 there was a difference in price of £6 a ton between soda ash and caustic and this gave consumers an additional incentive. Producers reacted by marketing liquid caustic soda. The crisis with the paper-makers came a little later, though it had been anticipated quite early in the career of U.A.C. As soon as the new combine was formed, Scottish paper-makers had planned their own chemical works, but nothing had happened. (*C.T.J.*, 3 Jan. 1891, p.4.) U.A.C. supplied both bleaching powder and caustic soda to the paper-makers at home and overseas. One blow to this business was the advance of wood pulp over other materials as a material for paper manufacture, for it required only sulphuric acid to bleach it. (Spicer, 1903, pp.206, 219, 220.) Competition in the supply of the paper trade came from electrolytic works—Swedish orders were lost in 1900, for instance, to the new Electro-Chemical Company of Bengtfors, Sweden, and in 1902 German works were attacking not only the Swiss but even the Scottish markets for alkali products. Trouble now also began to come from the paper-makers who had realized that by buying soda ash they could, like the soap firms, produce their own caustic soda and save on the operation. Until 1904, Young, Trotter and Son of Churnsides, Berwickshire, bought 1,000 tons of Allhusens high-strength caustic soda annually, nually, an amount ranging from 4 to 5 per cent of the works' capacity. They then decided to causticize purchased ammonia soda. The new unit at Churnsides cost £2,850, and although they had previously been sold their caustic needs at an exceptional price of only £9.15 a ton, concessions were now offered in an attempt to prevent them from building. Coal and lime were both costly at Churnsides, but they went ahead, and after covering standing charges on their own plant, were reckoned to save some £1.25 a ton. (U.A.C., Tyne.) U.A.C. soon faced similar challenges from others. The situation was complicated by Continental and American competition even in the British market. In 1894 the Solvay Process Company of Syracuse, New York State, reduced its prices, and competition became so keen that U.A.C. was at one time selling caustic in England at $10–$14 a ton more than in New York. The Solvay Process Company therefore 'at once invaded the domain of its competitors, once deemed impregnable'. The result was a visit to the United States of two prominent officers of the U.A.C., 'to confer with representatives of the home concern'. Prices were again raised. (Rothwell, 1895, p.99.)

The demise of U.A.C.

Gradually U.A.C. edged its way out of the alkali trade (Fig. 31). By the early twentieth century it was already a minor factor in soda ash. As its product range widened—by World War I Castner-Kellner were advertising 10 different products, Brunner Mond 15, but U.A.C. 87 (Davis, 1913, pp.56, 45, 247)—so sulphuric acid making became more obviously its central activity. By World War I, the trades in superphosphate and sulphate of ammonia were more important consumers of sulphuric acid than was the alkali trade. The wastefulness of labour and of fuel accelerated the decline of the remaining elements of the Leblanc process during the war; and in October 1920 the last Leblanc works in England, Allhusens, closed, though caustic production continued there. In 1920/1, during the depression, U.A.C. closed Mort Liddell works at Widnes and three of its remaining six works in St. Helens—Kurtz, Globe, and Greenbank. In 1925 four surviving caustic soda works made about 52,000 tons, or less than half the level of thirty years earlier. Finally, in 1926, together with Brunner Mond & Co. and its subsidiaries (including Castner-Kellner), Nobel Industries, and British Dyestuffs, U.A.C. became part of I.C.I. In the next four years seven more of the few surviving U.A.C. works were closed, including the last works on the Tyne, Tennants of

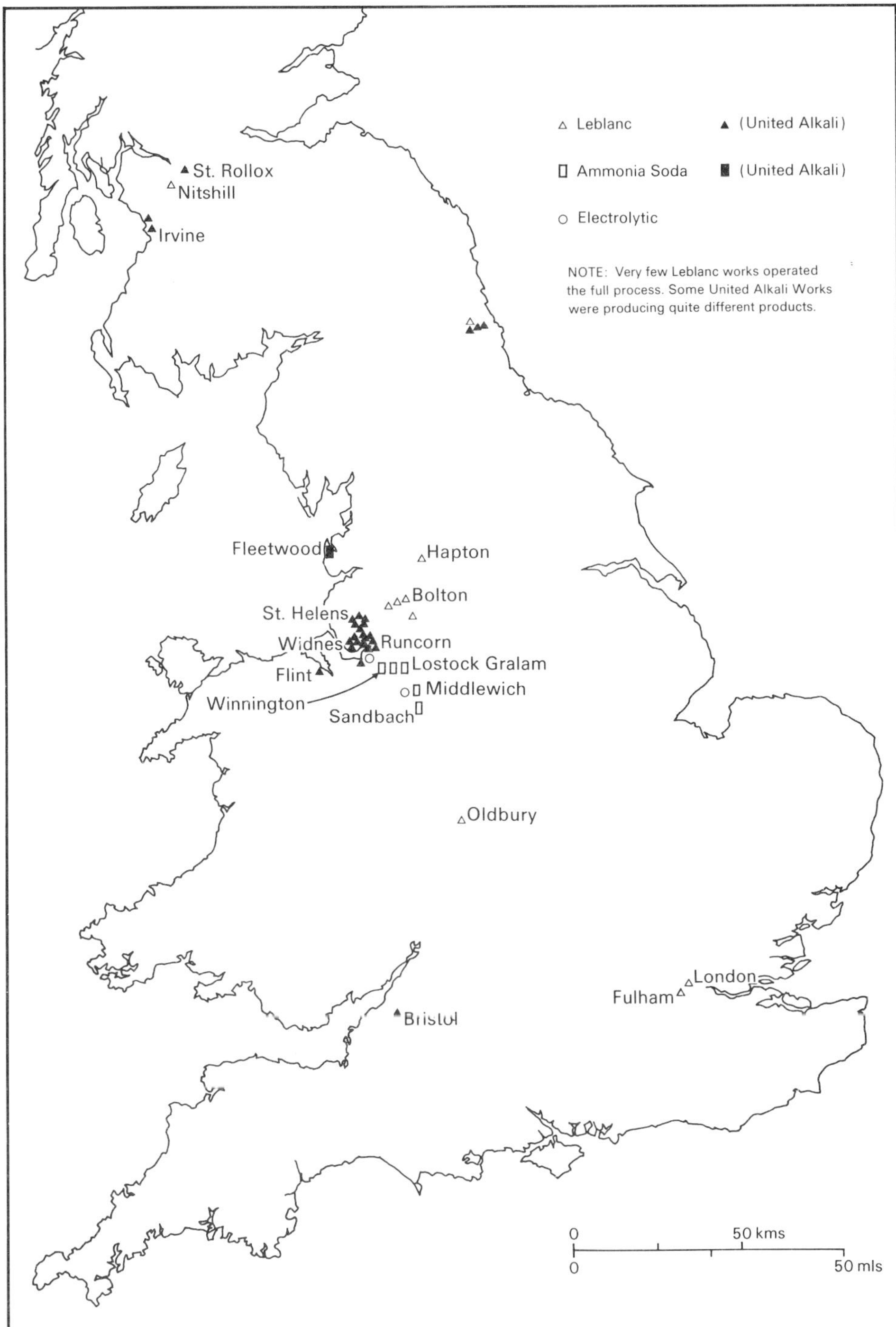

Fig. 31. British alkali works 1913 *Source:* Davis, 1913

Hebburn in 1931 and Allhusens in 1932. Baxters and Hardshaw works at St. Helens, the Clyde Wharf works, the Weston works, and the Tharsis Copper Works in Widnes also went. Among the seven plants to survive were Wigg, Netham, and Fleetwood Ammonia Soda works; the other four, some of them combinations of older works, were in Widnes. (U.A.C., 1930/2.) It was indicative of how far U.A.C. had come from its earlier specialization that, when I.C.I. renamed its constituent parts, the descendants of Brunner Mond became the Alkali Division and the remnants of U.A.C. were renamed General Chemical Division. It was perhaps typical of the ethos of U.A.C. that as late as 1926, a month before the formation of I.C.I., Sir Max Muspratt, U.A.C.'s chairman, was opening new offices in Gateshead for Allhusens.

CHAPTER 15

Electrolytic Processes

In autumn 1884 James C. Stevenson, of Jarrow and Friars Goose Chemical Works, surveyed current conditions and trends in the trade: 'The inevitable revolution is progressing which will leave to the ammonia process the production of soda ash and confine the Leblanc process to the production of caustic soda, crystals and bicarbonate' (*J.S.C.I.*, 29 Oct. 1884, p.503). He might well have added bleaching powder, for this proved a most important source of income to the old alkali firms during the eighties. In 1863 British bleaching powder production had been 30,000 tons; by 1894 output was 150,000 tons. In the mid-nineties the transfer of electrolytic processes of alkali manufacture from the realms of theory or of the laboratory to commercial levels of production undermined the redoubts of Leblanc defence in both caustic soda and bleaching powder. With electrolysis the only two important raw material inputs were brine and coal (for steam-fired electricity generation); there were no problems of bulky waste material.

As early as 1834, in his *Experimental Researches in Electricity*, Faraday had described the production of chlorine at one electrode and sodium at the other in the electrolysis of brine (Hardie, 1950, p.153). Charles Watt took out a patent for the process in 1852, but the first experiments on an industrial scale were not made for another thirty years. There were technical and commercial difficulties to overcome. The technical difficulty was that the chlorine produced at the anode reacted at once with the sodium hydrate to produce sodium chloride again. It was eventually shown that this could be prevented by the interpolation of a diaphragm between the anode and cathode or alternatively by the use of a so-called mercurial cathode in which the bottom of the trough in which the reaction was to take place was covered with mercury. Sodium once formed dissolved in this and so was protected from reaction with chlorine. After its removal from the trough, the sodium was extracted from the mercury. Once the practical problems of electrolysis had been solved, commercial success awaited low-cost electricity supply, and therefore the development of an efficient dynamo, following work by Siemens in the mid-1860s.

The first commercial operations date from the formation of the Chemische Fabrik Elektron A.G. and the construction by it of a 200-horsepower plant at Griesheim near Frankfurt in 1889/90. British experiments began in 1891/2, and in the following year the Hargreaves-Bird (Diaphragm) Cell was patented, largely by the prolific Widnes chemical engineer James Hargreaves. In 1892 Hamilton Castner patented his mercury cell. In the next few years a new British electrolytic alkali industry began to emerge, though success was achieved only after serious difficulties had been overcome.

Energy and labour costs were both reduced by the electrolytic process. The Leblanc producer of caustic soda had to provide labour for the loading and unloading of saltcake and black ash furnaces and for handling waste; the electrolytic alkali producer dealt with liquids and gases right up to the point at which the sodium hydrate solution was boiled to produce solid caustic soda. As Blount (1896, pp.819–20) observed,

> In these days when the manual worker is exigent, any process which tends towards his deletion has a commercial advantage over those of its competitors which depend upon his aid. It is not merely a diminution of running expenses which may be secured; it is the freedom from the fear of a disastrous stoppage of the whole works which makes labour-saving methods and appliances attractive to the manufacturer.

On the other hand, the electrolytic process required a larger capital outlay than the Leblanc process, and hence made high levels of operation more imperative. Blount (1896, p.820) summarized its advantages and weaknesses:

> The electrolytic process is direct, cleanly, economical of labour, free from worthless by-products. Per contra, its units of plant are small, troublesome, costly and quickly deteriorated by use. The chemical process can provide a large output with comparatively few units of plant, of, on the whole simple construction and of large size; repairs though heavy are not numerous, nor have they to be applied to a vast number of small apparatus. On the other hand, much arduous and fairly skilled labour is requisite, and by-products valueless or of negative value are invariable concomitants.

Given inevitable technical problems in the early years, the uncertainty over prices in the first decades of the electrolytic alkali industry created serious difficulties. Prices for bleaching powder were much more volatile than those for soda ash, and after the first big electrolytic projects were launched in 1895 and again after 1901 there were sharp falls (Table 64). Many doubted the ability of the new firms to compete, and experience of the early years went far to confirm their doubts. There were some optimists, but a strong undertone of doubt persisted. (*C.T.J.*, 24 Feb. 1896, p.548.) Kershaw (1897, p.302) stressed that margins were narrower than once had been anticipated. Nine years later he still had reservations (Kershaw, 1906, p.343) and regarded an even wider product range as the prerequisite of success for the electrolytic processes: 'Bleaching powder and caustic or carbonate of soda [i.e. soda ash] will no longer provide large profit for the makers by the Leblanc or electrolytic processes and it is to the by-products and special manufactures of each type of works that one must look for the dividends of the future.'

Growth and location in electrolytic alkali

The Leblanc process had been generally coal-oriented in location. Ammonia soda production was strongly localized on the brine-fields. The new process presented other locational considerations. Electrolysis was above all power-intensive. The question soon arose whether this power would be coal-based or

Table 64. *Average prices for bleaching powder and soda ash (58%) in December*

	Bleaching powder			Soda ash		
	£	*s.*	*d.*	£	*s.*	*d.*
1895	7	0	0	3	17	6
1896	6	12	6	3	10	0
1897	6	2	6	3	17	6
1898	4	13	6	4	0	0
1899	6	0	0	4	7	6
1900	6	17	6	4	12	6
1901	6	17	6	4	17	6
1902	6	5	0	4	10	0
1903	4	0	0	4	10	0
1904	4	1	6	4	10	0
1905	4	10	0	4	10	0

Source: Kershaw, 1906.

hydroelectric. Swan (1898, p.128) noted early on that so far steam-generated electricity had kept the industry in Lancashire but wondered whether it would eventually move to new 'habitats'. In Continental Europe and in the United States, where big power sites were available, this did happen—as Witt (1909, p.54) romanticized: 'Norway, in whose rocky solitudes the bear was wont to ramble and the elk and reindeer to graze, whose blue fjords knew no other craft than fishing-smacks and occasional pleasure yachts, begins to develop a chemical industry of vast dimensions.' In Britain search for cheap power had taken the aluminium reduction industry to Foyers by 1895, but in alkali the intensity of power use was less, and nearness to customers and interconnections with pre-existing industry in the same trade greater. Lunge (1910) recognized both the ideal and the practical constraints:

> It stands to reason that the electrolytic processes have been principally developed in localities where the electric current can be produced in the cheapest possible manner by means of water power, but this is not the only condition to be considered, as the question of freight to a centre of consumption and other circumstances may also play an important part. Where coal is very cheap indeed and the other conditions are favourable, it is possible to establish such an industry with a prospect of commercial success, even when the electric current is produced by means of steam engines.

Given the proximity of coal and salt on Merseyside or Teesside, the fuel economy as compared with Leblanc's process, and the need for brine rather than dry salt, a brine-field or intermediate location would have seemed ideal. However, like all new industries, the electrolytic alkali industry had to grope its way to locational rationality. Because only a few units were involved, and there were serious difficulties with the process, it is not surprising that what were inherently the best locations did not necessarily succeed (Fig.32).

Although the *North East Daily Gazette* of Middlesbrough hailed the electrolytic process as potentially of an importance second only to that of the discovery of the Cleveland Main Seam (quoted *C.N.*, 4 Nov. 1892, pp.233, 234), no plant was built on Teesside and the early projects were mostly in the

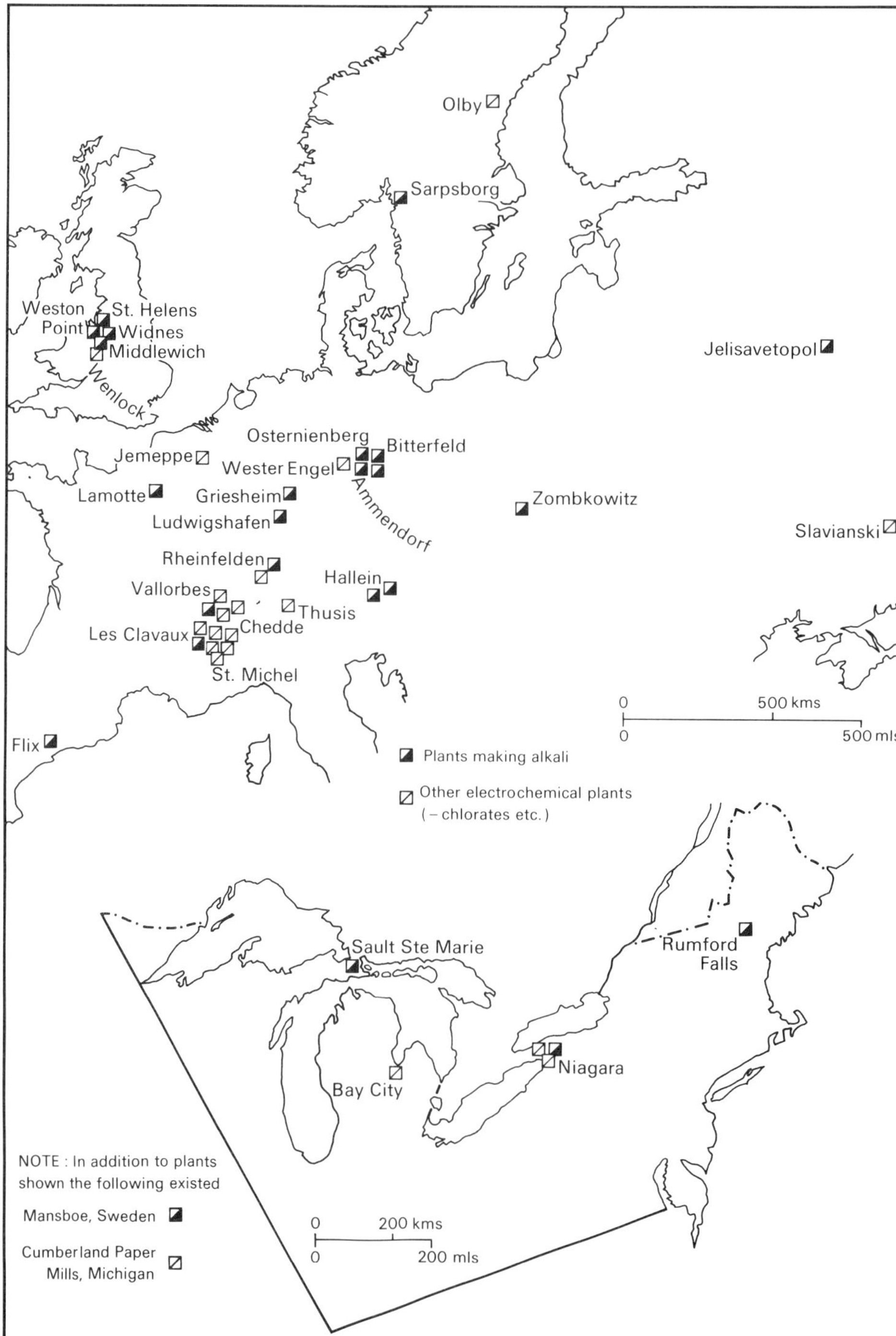

Fig. 32. Electrolytic plants 1900 *Source: C.T.J.*, 20 Jan. 1900

Mersey Basin. Some never got beyond the experimental or even conjectural stage, such as units to work the Vautin process at Bolton, a Parker process plant at Wednesbury in the Black Country, U.A.C. experiments at Widnes, and rumours of work on the process by Brunner Mond. (*C.T.J.*, 16 Oct. 1894.)

The British pioneer was the Alkali Chemical Company which carried out research into the process in 1891/2 and then produced electrolytic alkali for two years just south of Rochester at Snodland on the Medway where there was a large paper-mill. In 1894 the Electro-Chemical Company was formed with a capital of £0.2 million to work the Snodland process commercially. Several sites near coal, salt, or limestone were offered to them in North Wales, Cheshire, and Lancashire. After careful inspection, decision was taken to build in St. Helens on the site of the old Parr Copper Works. St. Helens had the disadvantage that rock-salt had to be brought in and then dissolved to produce the necessary brine; but it offered some advantages. There were associated trades, labour used to alkali production, though by very different methods, and management skills. (Surprisingly Electro-Chemical appointed as their manager an old Leblanc man (Barker & Harris, 1954, 444–5).) St. Helens offered environmental advantages as well: as the chairman pointed out, in a mainly agricultural area like that around Snodland, 'though no chlorine is allowed to escape, an injunction might have been applied for, entailing legal expense' (*C.T.J.*, 23 Nov. 1895, p.325).

After initial difficulties with machinery, the Electro-Chemical Company installed what were optimistically called 'excellent' engines and dynamos, and in 1894 they claimed that they would make caustic soda at 50 per cent below market prices; a little later, even more optimistically, they put their expected costs at 50 per cent below that of older methods of production (*C.T.J.*, 16 Feb. 1894, p.388; 24 Feb. 1894, p.125). Early operating experience moderated these hopes. By late 1897 weekly production was 130 tons of bleaching powder, 70 tons of caustic soda, and 5 or 6 tons of chlorate of potash; this at current prices left them, in their chairman's words, 'a fair profit' (*C.T.J.*, 27 Nov. 1897, p.341). However, prices fell sharply in 1898 and the firm never made the intended extension of their power plant from 3,500 to 6,000 horse-power. Prices moved up in 1899, but there were further difficulties and expectations that more capital might be needed. After several short stoppages, the works was closed. (*C.T.J.*, 20 Jan. 1900, p.38.) It was recognized as needing major outlay for new equipment and had made big losses even though prices for bleach and caustic soda were high. In 1901 it was decided to wind up the company and sell the works. By 1904 the company was liquidated. (*C.T.J.*, 2 Feb. 1901, pp.107–8, 16 Feb. 1901, 9 Mar. 1901, p.234; Barker & Harris, 1954, pp.444–5.)

Apart from an abortive scheme in the Shropshire coalfield, the only other early British electrolytic alkali projects were on or near to the Cheshire salt-field. In 1898 the Salt Union made 'industrial scale' trials for alkali and chlorine compounds production at Winsford but gave up when they proved unsuccessful (*C. T.J.*, 20 Jan. 1900, p.38). Hargreaves's process was carried on at Farnworth by the General Electric Parent Company from 1895 for some years. This willingness to continue experimental production and meanwhile to survey the progress of other more precipitous entrants to the production field, along with Hargreaves's reputation, and the relatively modest cost savings which were claimed, commanded respect. By late 1897 the company was convinced of the quality of the process and began to sell the rights. In 1899 the Electrolytic Alkali Company was formed with a capital of £0.5 million (by far the largest capital for any electro-chemical works in Britain). By spring 1900 works were approaching completion at Cledford Bridge on the Trent and Mersey Canal just east of Middlewich. As well as canal access, the site had good rail facilities, and ample brine supplies which were free of royalties. (*C.T.J.*, 7 Dec. 1895; 1 May 1897, p.299; 18 Feb. 1899, p.122; 20 Jan. 1900, p.38; 14 Apr. 1900, p.252.)

The commercial press remained sceptical. A year after completion, when commenting on the declaration of a 2 per cent half-year's dividend by another electro-chemical company (Castner-Kellner), the *Chemical Trade Journal* (8 June 1901, pp.521–2) noted that all was far from well with the electrolytic processes. Prices were not too good, there was surplus production, and, at the same time, economy in use by important customers, notably the paper-makers. 'We await with much interest the first report of the Middlewich Electrolytic Plant, and, in passing, concluded with the remark that with bleach and caustic at present prices, a well-managed Leblanc works, erected on the best system, and supplied with ample working capital, would pay a dividend nearer 20 percent than 2.' The Electrolytic Alkali Company was liquidated in 1913. In the same year it was taken over by a new concern, Electro Bleach and By-Products Limited, which in turn was absorbed by Brunner Mond in 1921. Six years later their works was shut down and dismantled. (Hardie, 1950, p.197.)

The emergence of the Castner-Kellner Company

In 1886 Hamilton Castner, a young American engineer, came to England and became involved with aluminium manufacture at Oldbury. By 1892 he had patented an electrolytic cell, and two years later an electrolytic alkali plant was put up at Oldbury. Castner's cell became linked with the innovations which the Austrian Carl Kellner had made in the development of the mercurial cathode cell. It was not immediately obvious that the Castner-Kellner process was superior to others. (*C. T.J.*, 23 Nov. 1895, p.325.) In October 1895 the prospectus of the new Castner-Kellner Alkali Company was issued. Its capital was half as great again as Electro-Chemical's. Recognizing the importance of access to brine, good home outlets, and easy transport to foreign markets, the company looked to the North-West. It was helped in this by a distinguished board of business and scientific persons from that region, including the chairman, Sir William Mather of the engineers Mather and Platt of Oldham who built the works, Henry Roscoe the professor of chemistry at Owens College, and, significant from the point of view of marketing, Frank Hardcastle, President of the United Bleachers Association of Lancashire and Cheshire. It was (wrongly) believed at the time that although their name did not appear on the board, Brunner Mond & Co. were associated. Weston Point, excellently placed on the newly opened ship canal for access to Liverpool docks, and where large amounts of brine (up to 10,000 gallons daily) were made available from the Salt Union's Runcorn reservoir, was the location chosen for the new works (There were, however, disadvantages: road access was poor until after World War I, and the branch railway from Runcorn was not built until 1922 (Smith, 1953, p.258).) The site cost only £5,500. By April 1897 £70,000 had been spent on the plant, and it came into production that year. Castner-Kellner held advanced ideas in regard to working conditions and environment. These were spelt out at the First Annual General Meeting (*C.T.J.*, 3 April 1897, pp.238–9). As theirs was going to be 'a lasting business' they had 'determined to put up works such as had never been erected for chemical purposes in England or elsewhere before'. William Mather stated that the Castner-Kellner Company had taken care to give their men every convenience in clothing and so on—'He need not say that no profit in the world could be, or ought to be acceptable to any right-minded man that was earned at the well-being of any other man.' Of the environment he said:

> Their works were situated in the rural part of the country. They saw the fearful shroud overhanging the district known as Widnes, and were cognizant of the prevailing odours which sprung from the escape of chlorine and other gases. They had from the start endeavoured to make their works harmless as regards vegetation and as regards interfering with the convenience of any of the residents in their neighbourhood. A certain expense had been incurred in absorbing those gases which otherwise were always sent up the chimney and spread far and wide.

Soon afterwards Mather had to apologize in the technical press for his remarks and explain that he did not mean to imply that Widnes was still as bad as it had been in the past, and to say that he would be sorry if his remarks about the United Alkali Company were 'construed to impugn the character of that great company for the utmost solicitude in the public welfare in carrying on their manufacturing operations' (*C.T.J.*, 10 Apr. 1897, p.253).

From the start there was keen competition with U.A.C. and some uncertainty about the outcome (Table 65). It was necessary to persuade customers to buy the new product, a purer caustic soda than was produced by the chemical process—the same sort of consumer resistance had faced Muspratt in Liverpool seventy years earlier, or Brunner Mond & Co. when they tried to market their higher-strength soda ash in the mid-1870s. Fuel prices rose, demand was uncertain, and prices see-sawed. Dividends in the early years were sometimes as high as 8 per cent but often no more than half that. Weston Point came into production when bleaching powder prices were falling, and its advent caused further price-cutting. In September 1897 bleaching powder was £5. 18*s*. 6*d*. a ton, but sometimes by the end of the year it could be had for as little as £4. 15*s*. or even less. In this fierce fight with the Leblanc process, Castner-Kellner were helped by the ammonia soda interests. Commenting on the low figure of £4. 15*s*. a ton, the *Chemical Trade Journal* (1 Jan. 1898, p.1) noted, 'This is not a genuine manufacturing figure, however, as the whole output from Weston Point has been purchased by the Solvay Syndicate at about £5. 17*s*. 6*d*. and it is this syndicate that is so philanthropically giving paper makers and bleachers their requirements at less than cost price.' It had been done, as John Brock told his U.A.C. shareholders in March 1899, 'for the express purpose, as they put it, of "keeping us in order"' (*The Times* 18 Mar. 1899, p.16). During this period the Leblanc makers continued to hope for something to turn up to help them. The Tyne Directors of U.A.C. resolved to keep some Tyneside plant in existence, though idle, 'as in the event of a collapse in the process of decomposing salt by electricity it is the opinion of the Local Directors that increased demand for bleaching powder from the continent of Europe would again arise' (U.A.C., Tyne, 16 May 1898). Difficulties of various kinds did indeed continue to trouble the electrolytic makers.

Table 65. *Costs and prices for alkali products, 1893*

	U.A.C. estimated production costs £ *s.*	*Selling prices* £ *s.*
Soda ash 48%	5 0	3 0
Caustic soda	8 0	8 10
Bleaching powder:		
by Deacon process	3 15	8 10 to 10 0
by Hargreaves/Robinson process	3 5	

Source: Rothwell, 1893, vol. 2, p.123.

In 1899, when demand for caustic soda was boosted by the introduction of 'mercerizing', a process which gave an attractive silky appearance to cottons and required 300 tons of caustic soda weekly for Lancashire and Yorkshire use, there were fears that this high level of production would lead to surplus chlorine production and therefore a collapse of bleaching powder prices. Moreover, in so far as they produced chlorine in the form of hydrochloric acid, United Alkali seemed to have a trump card because of the outlet for this acid in the galvanizing trade. This, however, proved not to be as significant as the electrolytic interest feared. None the less, U.A.C. had an enviable flexibility: it had plant which had been written off, but which could be re-activated when demand was high, and a so-called 'healthy rivalry' between its constituent works had resulted in a sharp reduction in its production costs.

(*C.T.J.*, 2 Nov. 1895, pp.271–3, 22 May 1897, p.348, 8 Dec. 1899, pp.730–1; *Eng.* 14 Mar. 1902, p.352.)

Yet in spite of all the uncertainties of the trade, Castner-Kellner were soon successfully established. They quickly obtained contracts for the sale of all the first three years' output. Within six months of beginning work they were so satisfied with their level of costs that, with the second production unit already almost complete, decision had been taken to install two more units. (*C.T.J.*, 9 Oct. 1897, p.230.) Their technical success is indicated by the fact that, of thirty-five electrolytic alkali works in production or nearly ready to begin outside Britain by the beginning of 1900, seven were using the Castner-Kellner process (*C.T.J.*, 20 Jan. 1900, pp.38–40). It proved competitive. In 1902 bleaching powder prices were almost £7 a ton, and during that year co-operation between British and German makers ended and prices fell rapidly. By December some contracts had been placed for 1903 at as low as £4. A year later this was the ruling price. Castner-Kellner were less affected by the fall than the Leblanc industry, and it was said that their production costs were as low as £3. 15*s*. a ton. (*Eng.*, 12 Dec. 1902, p.790; Kershaw, 1906.) Uncertainties in the trade certainly hit the Leblanc firms harder—thus in 1898 bleaching powder cost the Widnes Alkali Works 58*s*. 3*d*. a ton to make; the selling price was then 96*s*. 6*d*. In the first half of 1908 the selling price was 75*s*. 9*d*. and costs of production were now 92*s*. 5*d*.

Within a few years Castner-Kellner were involved in locational change. For a number of years they sent caustic soda to Oldbury where sodium metal was made. In 1901 Castner-Kellner absorbed their business parent, closed Oldbury works, and moved its production of sodium and sodium peroxide to Weston Point. However scope for expansion there proved inadequate, and it was quickly decided to move sodium production yet again. Power supply was an important consideration, and the large capacity of the Newcastle Electric Supply Company was attractive, and Castner-Kellner chose to locate their plant at Wallsend. An attempt was made to use Allhusens' caustic soda at this works, but it proved unsatisfactory—presumably partly because it was of lower purity than electrolytic caustic—and Wallsend continued to obtain its caustic from Weston Point. (Castner-Kellner, 1945.) By the 1920s Allhusens had an electrolytic alkali unit at Gateshead, supplied with brine from Haverton Hill by rail tank car, but it is not clear whether this ever supplied Wallsend. The Gateshead electrolytic works was closed after the formation of I.C.I. (Pilbin, 1935, pp.309–10.) Later hopes that Tyneside might become a major centre of electrolytic alkali manufacture (*Industrial Chemist*, Feb. 1927) were first frustrated by the Billingham project and then dashed completely when in 1929 Wallsend was closed and operations were concentrated at the Cassel Works on the Billingham site.

In 1914 United Alkali built an electrolytic alkali plant at Sullivan works, Widnes. There were bigger plans which never came to fruition for works on either the Cheshire or the Tees salt-fields. (Hardie, 1950, p.153.) Castner-

Kellner succeeded and expanded (Table 66) until, in 1920, Brunner Mond & Co. acquired a majority share. By the mid-1920s the independent merchant electrolytic alkali trade had ceased to exist.

Table 66. *Castner-Kellner capacity for bleaching powder and caustic soda* (tons per week)

1897/8	400
1914	1000
1920	1800

Source: Record, 1926

CHAPTER 16

The International Framework: markets and competition

From the early nineteenth century the alkali industry worldwide was a growth industry. Soda ash production reached an estimated 0.3 million tons in 1863, had passed 1 million tons by 1890, and 1.5 million by 1900. By 1913 output was 2.8 million tons and by the time of the formation of I.C.I. in 1926 had reached 4 million. (Hou, 1942, p.41.) Figures are not available for other alkali products, though it is clear that for some of them the rate of expansion was not as great. In the early years Britain almost monopolized production. As late as 1881/2, counting all alkali in terms of soda ash, Weldon estimated that the United Kingdom made 432 out of a world total of 709 thousand tons. Thereafter, however, the growth of British output lagged. Between 1882 and 1894 world production increased 50 per cent (to 1,051,000 tons): Britain's output went up by only 21 per cent. A more decisive indicator of decline was Britain's continuing use of the Leblanc process when others were abandoning it. In this twelve-year period world output of Leblanc soda products went down by about one-sixth from 545 to 456 thousand tons; in Britain from 380 to 340 thousand tons. There were other indicators of decline. Thus as late as 1900 the United Kingdom produced almost twice as much sulphuric acid as the German output of 0.55 million tons, but by 1913 Britain's output was only 1.1 million tons to Germany's 1.7 million (Landes, 1965, p.501).

The slow growth and the technical backwardness of the British chemical industry caused comment and grave disquiet in many quarters. By 1901 the American *Engineering and Mining Journal* was writing, 'While less is said about it in public than of the iron and other trades, there is no important branch of industry in Great Britain that has suffered so much from foreign competition as the chemical manufacture' (quoted *C.T.J.*, 17 Aug. 1901, p.137). Few seem to have recognized that the Leblanc process was perhaps more suited to Britain's resources and industrial structure and stage of development, and that, naturally, as the rest of the world began to grow economically, the rate of expansion in Britain would be slower. Alfred Mond (1903, pp.217–18) viewed the changes with more equanimity and perception than most: 'To anyone who gave consideration to the matter, it must have been obvious that this development would take place. In fact it would have been impossible for a small country with a small population like England to supply the industrial demands of the modern civilized world entirely by itself'. That there might be advantages in a spread of industrialization from Britain few foresaw, though a continuing role as workshop to the world, even if commercially feasible, would have been environmentally disastrous. The majority, whose interests were confined to either the return on their capital or the income from their labour, deplored Britain's chemical decline. Explanations for the process of shift of growth elsewhere were varied but there were a few common themes. A leading one was the disturbing effect of tariff policies on 'natural' patterns of trade. Others, less popular, concerned unfavourable British response to innovation, technical education, business initiative, and labour attitudes. It is impossible to disentangle all the threads or even, perhaps, to reach a balanced assessment, but overall one can see the initiative coming from overseas in the form of a natural and strong desire to build up home industries, to be met by an inadequate—and, given her economic maturity—perhaps an inevitably inadequate response from Britain.

Tariffs and British overseas markets

The Leblanc alkali industry had grown up in a world relatively free of tariffs. Proposals for foreign protection had always caused concern to the industry. In the 1840s there was an important trade in soda crystals from the Tyne to Germany. At the end of that decade the Prussian government was proposing to raise the duty to £3. (*M.J.*, 7 Apr. 1849.) These threats were relatively small and short-lived and it was long believed that fuel and salt costs were too high for Germany to build up an alkali industry (Richardson & Watts, 1863, p.321). In 1860 the British alkali trade was helped by the reduction in duties associated with the Cobden Treaty with France; French output of alkali in 1865—mostly near Marseilles—was no higher than in 1853. In the late seventies, however, things began to go amiss with foreign trade, and from this there was no recovery (Fig.33).

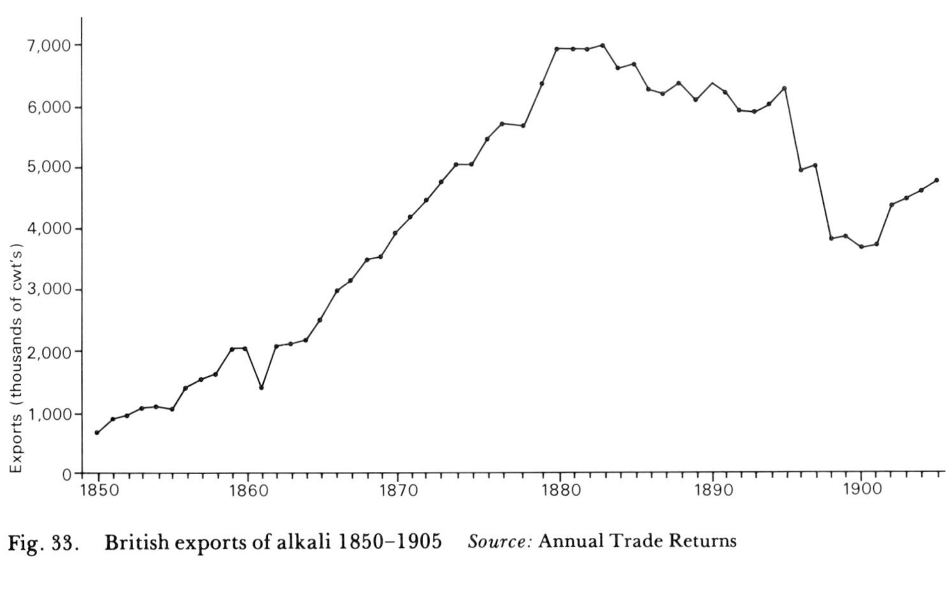

Fig. 33. British exports of alkali 1850–1905 *Source:* Annual Trade Returns

In 1873 duties on soda salts imported to Germany were cut and some parts of the German industry shrank. According to Hasenclever, one of the leaders of the German alkali trade, German alkali production from 1872 to 1878 fell from 58,000 tons to 42,000 tons as a result of British competition, though other sources suggest a smaller decline. (Haber, 1958, pp.47, 123; Landes, 1965, p.415.) In 1879 the tariff was reformed and the higher levels of the duties for soda ash, though not for other alkali products, coming, it must be noted, at the same time as the competition from the Solvay process, brought rapid expansion. In less than eight years to the mid-1880s German production of alkali almost tripled. Imports from Britain, formerly three to four times the level of German exports, dwindled and German exports became important (Fig.34 and Tables 67, 68). By 1884 Germany had become a net exporter of soda ash, in 1886 of soda crystals, and by 1890 of caustic soda. By the early 1890s only bicarbonate of soda was a net import. By this time over one-fifth of all German soda production was exported. (*J.S.C.I.*, 30 June 1891, pp.582–3.) The old plant was swept away. In 1880/1 there were twenty-one Leblanc works in Germany; by 1901 only five, and in 1909 two. (*J.S.C.I.*, 28 Feb. 1901, p.290; Martin, Smith & Milson, 1915, p.36.) Tariffs were often deplored by free traders as protecting home industries at the expense of home consumers, but in this instance the German consumers gained impressively from the changes. To 1877, soda ash prices in Germany were as high as 23 Mark per 100 kg, but by late 1884 only 12–14 Mark. (*J.S.C.I.*, 29 Dec. 1884, p.631.)

Table 67. *German production and trade in soda, 1877, 1883, 1890, 1894* (thousand tons)

	1877	*1883*	*1890*	*1894*
Production:	42.5	115	195	250
of which Leblanc	42.5	59	N.A.	40
Imports	27.5	N.A.	N.A.	N.A.
Excess of exports over imports	—	3.3	28.4	N.A.

Sources: *J.S.C.I.*, 31 Aug. 1893, p.726 and various.

Table 68. *United Kingdom and German exports of alkali, 1881, 1891* (thousand cwt)

	U.K.		*Germany*
	Total	of which to Germany	
1881	6947	929	351
1891	6226	144	868

Source: *Eng.*, 12 July 1901, pp.54–5.

Production in France grew faster than in Britain, but much slower than in Germany. Here too ammonia soda made rapid progress. France also had imposed protective duties, which were generally heavier than the German ones

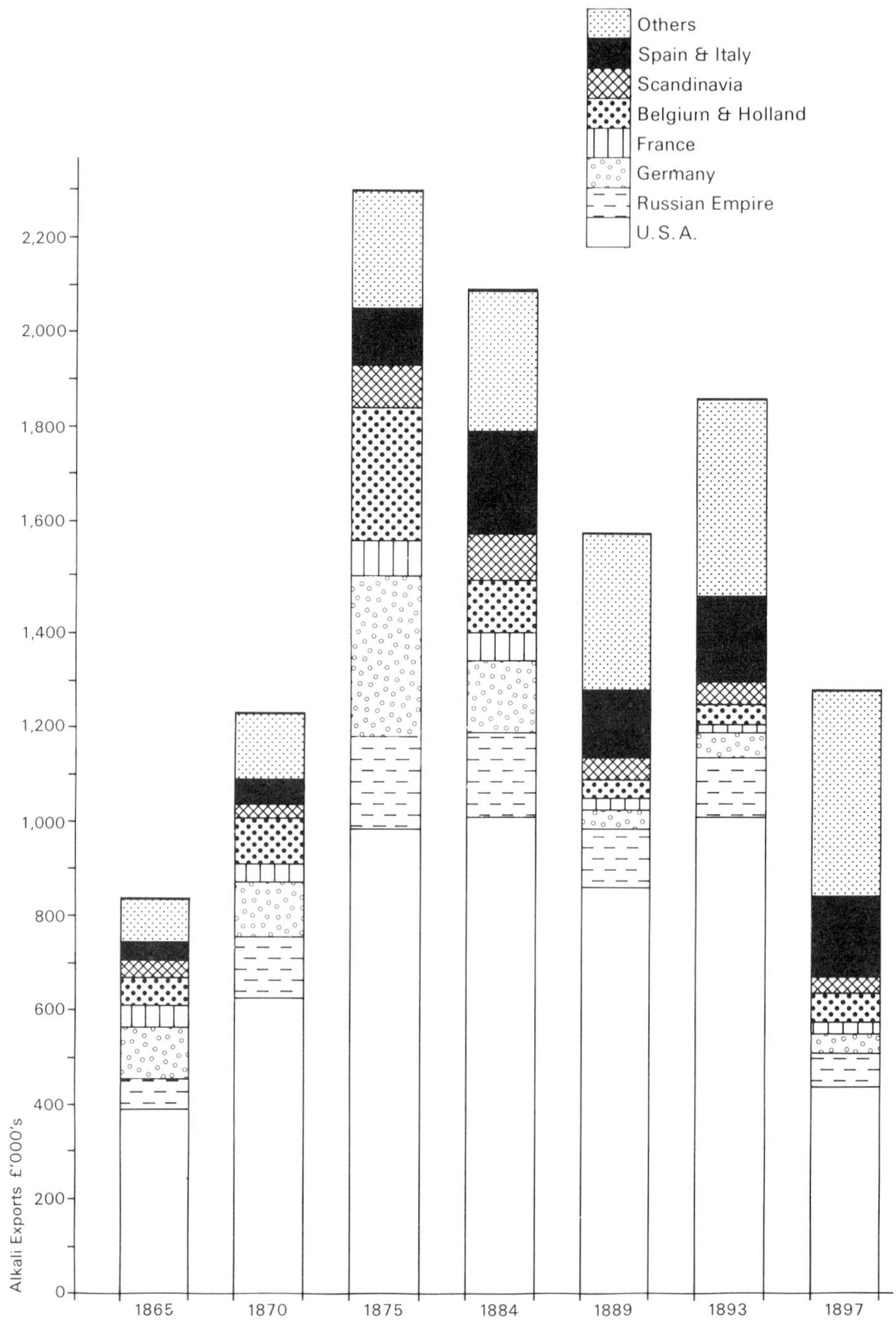

Fig. 34. British exports of alkali by destination 1865–97 *Source:* Annual Trade Returns

(Table 69). In 1882 Germany and France together made only 227,000 tons of alkali products to 432,000 in the United Kingdom, though their combined ammonia soda production was already almost twice as large (101,000 tons to 52,000). In the mid-eighties France was still largely dependent on British bleaching powder. By 1894 Germany and France turned out 420,000 tons of alkali products to the U.K. total of 521,000 tons, and at 360,000 tons their ammonia soda production was still twice as great. By the nineties, British exports of alkali to both countries were of very minor importance—under 10,000 tons a year to the German Empire and under 3,000 to France. In 1875/6 France was importing some 12,000 tons of soda annually, but by 1890 was exporting over 20,000 tons. (*C.T.J.*, 13 June 1891, p.380; Table 70.)

Table 69. *Import duties on alkali in Germany and France, 1882* (£ per ton)

	France	*Germany*
Soda ash	1.64	1.20
Soda crystals	0.76	0.75
Caustic soda	2.56	2.00
Bleaching powder	1.42	1.50

Source: Weldon, 1883, p.67.

Table 70. *Alkali production in France and Germany, 1860s–1901/2* (thousand metric tons)

	France	*Germany*
mid-1860s	45 (1865)	33 (1867)
1878/80	99 (1880)	42 (1878)
1891/3	146 (1893)	210 (1891)
1901/2	230 (1902)	325 (1901)

Source: Haber, 1958, pp.47, 110, 123.

Some countries of western Europe, Spain and Italy among them, remained until well after the end of the century dependent upon imported alkali. In other cases the old Leblanc industry was swept away and nothing, or inadequate capacity, took its place, so that imports had to increase. Thus the Netherlands became a much more important market for British alkali in the nineties than France, and did not build its own first ammonia soda works (at Roermond) until 1939. In Belgium conditions were not ideal for the Solvay process. In the nineties the Leblanc process was even operated in Belgium again, Lorraine rock-salt or English salt from brine being used. English bleaching powder for long continued to dominate the market there, though competition from Germany was keen. (*C.T.J.*, 10 Aug. 1889, p.79; Table 71.) For a time, in the eighties, the Austro-Hungarian Empire seemed likely to build up an important new-style alkali industry behind a tariff wall. By 1883 four ammonia soda works were built, building, or planned, and as late as 1894 output was comparable with that of the United States. (Weldon, 1883, p.68; Rose, 1901, p.46.)

Table 71. *Production by leading countries of Leblanc and ammonia soda ash, 1884 and 1894* (thousand tons)

	1884		*1894*	
	Leblanc	Ammonia soda	Leblanc	Ammonia soda
United Kingdom	380	52	340	181
Germany	56.5	44	40	210
France	70	57	20	150
Austro-Hungary	39	1.0	20	75
Russian Empire	—	—	10	50
Belgium	—	8.0	6	30
U.S.A.	—	1.1	20	80
Total	545	163	456	676

Source: *B.P.P.*, 1900, LXXV, p.46.

Britain retained an important role both in the Empire and in the rest of the developing world. Canada, Australia, and India installed their first ammonia soda works as late as 1917, 1940, and 1941 respectively. By 1897/8 the first two together were comparable as a market for British alkali to Italy. In Japan the only sulphuric acid works and small alkali works were at Osaka in 1883, and they operated under the considerable difficulties of a newly industrializing country. As a British journal put it, 'At present it would appear a matter of great difficulty to found a new chemical industry in Japan, owing to the lack both of capital and skilled labour, as also to the great obstacles in the way of transport of raw materials from different parts of the country.' (*J.S.C.I.*, 1883, p.21.) By 1897 the Osaka Alkali Company had to give up caustic production because it could not compete with imported material, and in 1898/9, though Japan was self-sufficient in bleaching powder, it still provided only one-third of its caustic soda requirements, the rest being English. (*C.T.J.*, 20 Jan. 1894, p.57; 28 Jan. 1899, p.60.) In 1916 and 1918 Japan's first two ammonia soda plants were built.

The Russian Empire and the United States

By 1890 the United States was the leading market for British alkali, and the Russian Empire ranked high. It was the Russians who made the earlier important steps to independence. Until the early 1880s there was no alkali industry in the Russian Empire, consumption being met by British alkali or by import from Belgium or Germany. Imports of soda ash began to fall off in the middle 1880s when the first Russian Solvay works was built. Approximately three-quarters of the alkali consumed in the Russian Empire was imported in 1887; by the mid-nineties consumption had increased by 50 per cent, but imports were negligible (Table 72).

By this time tariff arrangements with Germany enabled her to compete more effectively with Britain in a rapidly dwindling Russian import trade. The imperial government was reported to be revising railway rates for soda ash in the light of assurances from the sole indigenous ammonia soda producer, Lubimov, Solvay and Co., that it could make sufficient soda to supply the

Table 72. *Production, consumption, and importation of alkali into the Russian Empire, 1887–1900* (thousand metric tons)

	Production	*Consumption*	*Importation*
1887	11	*c*. 40	32 (1885)
1890	14	44	30
1895	47	64	5
1900	87 (1901)	80	3 (1899)

Source: Haber, 1958.

whole empire. (*C.T.J.*, 31 Mar. 1894, p.212; *J.S.C.I.*, 30 July 1892, p.649.) The Solvay producers made good their claim.

By about 1860, 30 per cent of the U.K. production of bleaching powder and as much as 60 per cent of soda production went to the United States. Glasgow and Liverpool were the chief exporting points, the prices of the latter dominating. (Hardie & Pratt, 1966, p.37.) Up to a late stage it was believed that high-cost labour ruled out the United States as a major chemical producer (Allhusen, 1886, p.268). The American market became progressively more important to the British industry as continental markets were cut off after the mid-1870s, and by 1893 it could be claimed that 'the American trade has grown to be the prop and mainstay of the United Alkali Company of Great Britain' (Wyatt, 1893, p.57; Tables 73, 74). The prop, however, was soon removed. In absolute terms alkali deliveries from Britain rose from £756,000 in 1870 to £1,257,000 in 1890, but by 1905 were as little as £74,000. In the early stages of the decline, this was not due to tariff policy but to the growth of a Solvay industry in the United States.

Table 73. *United States imports of soda ash and caustic soda, 1867–1892* (thousand dollars)

	Soda ash and sal soda	*Caustic soda*
1867	2736	549
1872	3119	1037
1877	3473	1043
1882	4009	1336
1887	2843	1651
1892	4699	1337

Source: Wyatt, 1893, p.66.

Table 74. *The United States as a market for British alkali* (percentage of value of all British alkali exports)

1870	51.5	1890	60.1
1875	42.6	1895	49.4
1880	50.2	1900	12.2
1885	56.5	1905	5.1

Source: Haber, 1958, p.214.

The Solvay Process Company works at Syracuse started production early in 1884. The following year it made 15,000 tons of soda ash. By 1888 Syracuse supplied one-third of the United States' demand for soda ash, but 125,000 tons were imported from Britain. (*J.S.C.I.*, 30 June 1890, p.672.) In 1893 the Wyandotte works of the Michigan Alkali Company and in 1894 the Mathieson Alkali works at Saltville, Virginia, came into production. Imports followed an irregular course, but in the four years 1890–3 reached their record levels. After 1895 there was a rapid decline. This was accelerated by the Dingley tariff of 1897, which sharply increased all duties on alkali; it was, however, indicative of the prime role of the increased indigenous production and the secondary role of the tariff that the rate of duty for crystals and caustic soda was still below that of a few years earlier (Tables 75 and 76). New plants and extensions to old ones boosted home production as the difficulties of foreign suppliers increased. By 1900 United States production was ten times greater than its importation, and in that year the United States produced 363,000 tons of ammonia soda ash or 28 per cent of the world's total. (Hou, 1942, p.32; Pennock, 1903, p.670.)

Table 75. *United States alkali tariffs, 1872–1907* (cents per lb.)

	Soda ash	*Soda crystals*	*Caustic soda*
1872	0.5	0.5	–
1878	0.25	0.25	1.50
1883	0.25	0.25	1.00
1890	0.25	0.25	1.00
1894	0.25	0.125	0.50
1897	0.375	0.200	0.75
1907	0.25	0.25	0.50

Sources: various.

Table 76. *United Kingdom exports of alkali products to the U.S.A., 1890–1904* (thousand tons)

	Soda ash	*Caustic soda*	*Soda crystals*
1890	145	41	17
		McKinley Tariff Act	
1892	156	27	15
1894	135	25	15
1896	83	24	10
		Dingley Tariff Act	
1898	32	13	4
1900	34	7	3
1902	17	4	4
1904	13	5	4

Source: *The Times*, 26 Nov. 1906, p.7.

Brunner Mond and United Alkali reactions

All British alkali exporters suffered severely from the changes in American business. British exports of soda ash to the United States had fallen from 126,000 to 29,000 tons between 1895 and 1898 and for caustic soda the fall had been from 34,000 to 11,000 tons. (*The Times,* 18 Mar. 1899, p.16.) For more than twenty years after they had entered the trade, Brunner Mond & Co.

had exported much more than they sold at home, and the United States was their chief market. From 1886 to 1898 about half of Brunners' output was sold to the United States. (Reader, 1970, p.97; Table 77.) They were however shareholders in the Syracuse works. Syracuse quickly reduced the imports—by 1887 almost 25 per cent of the American supply of alkali came from Syracuse,

Table 77. *Brunner Mond & Co. deliveries to home and export markets, 1880–1900* (thousand tons)

	United Kingdom	*Overseas*
1880	3.7	15.1
1885	20.0	52.3
1890	43.4	78.3
1895	68.5	90.9
1900	103.8	53.1

Source: Hill, 1964, p.185.

Brunners provided 33 per cent and the Leblanc firms about 42.5 per cent. (Reader, 1970, p.98.) In the first half of November 1896, before the Dingley tariff, they had already laid off 170 men at Winnington (*C.T.J.*, 7 Nov. 1896, p.294; 14 Nov. 1896, p.310). Brunner Mond & Co.'s 1898 production of soda ash was only 2,000 tons more than in 1890, whereas the increase from 1885 to 1890 had been one of 102,000 tons. Although Brunners claimed to have recovered from the blow to the American trade by March 1898, their growing difficulties in American markets caused them to look elsewhere, particularly to the Far East (Table 78).

Table 78. *British soda ash deliveries to eastern markets, 1901–1925* (thousand tons)

	Total exports all markets	*China*	*Japan*	*Australia*	*India*
1901	63.0	6.3	5.9	5.4	2.2
1905	75.2	11.7	8.8	5.9	6.4
1910	108.5	17.9	16.7	9.3	11.8
1915	216.2	31.4	25.8	15.5	25.8
1920	171.3	27.8	38.0	33.5	26.8
1925	230.6	47.6	75.9	25.8	36.8

Source: Hill, 1964, p.186.

In 1899 Brunner Mond & Co. sent Henry Glendinning on a two-year market exploratory mission, mainly in China. By 1905 Brunner Mond sold only 10,000 tons of alkali there, but by 1915 37,000 tons. (Little, 1915/16; Dick, 1973, p.65.) By the 1920s Brunner Mond had a network of Chinese agents throughout the country. (*Brunner Mond Record*, July 1926, pp.25–6.) Although Solvay works had pushed the British alkali firms out of the markets of European Russia, in the far-eastern regions the latter were able to compete successfully. By World War I the Piankov Glass works, the Primorski cement works, and five eastern Siberian soap-works were using Brunners' alkali. (Little, 1915/16.) The British Empire remained a major and for long fairly secure outlet.

U.A.C. suffered even more severely than Brunner Mond from the collapse of the United States trade. In 1894 they were selling caustic soda in the United States at such absurdly low prices that some of their American customers had made profits by sending some of it back to the United Kingdom for sale there (*C.T.J.*, 13 Oct. 1894, p.221). In November 1896 U.A.C. announced a cut in the chemical workers' week at Widnes from 56 or 60 hours to as little as 40 hours until the following February (*C.T.J.*, 21 Nov. 1896). The Dingley tariff which came into force on 1 October 1897 increased the duty on caustic soda to a level that was 70 per cent of the f.o.b. value and for soda ash 150 per cent. Late in 1897 it seemed likely that U.A.C. would build a major ammonia soda plant near Detroit; instead, smaller operations were established at Bay City further north. In spring 1899 John Brock assured his shareholders that 'with reasonable duties and fair play they could hold their own and hold it well'. In fact, there is a good deal of evidence to suggest that British plant and management were inferior.

British and foreign efficiency

In June 1881 Henry Roscoe of Owens College, Manchester, spoke on the theme of foreign competition to the new Society of Chemical Industry:

> We have heard much lately about foreign competititon, or even the decay of English manufactures. That foreign competition should become year by year keener, and that we who have led the way and had the advantage of the start in the race, should feel this competititon, can in no way surprise us. That English manufacturing skill is on the wane, or that the inventive genius of the people is dying out, I for one do not believe. Only let the Englishman know what he really needs, and he will soon take care that he gets it, be that need an increased technical or educational skill, or a wider diffusion of the knowledge of scientific principles. (Roscoe, 1881, p.1.)

Others were less sure. In 1880 Georg Lunge published his great work *A Theoretical and Practical Treatise on the Manufacture of Sulphuric Acid and Alkali, With the Collateral Branches*. The *Mining Journal* (25 Dec. 1880, p.1497), reviewing it under the subheading 'The Importance of Technical Education', wrote:

> The manufacture of chemicals has made enormous strides both in quantity and quality in France and even more so in Germany. Many of the chemicals of these countries outstrip those of English works in purity, and their plant and their processes are frequently superior to those used in the majority of English works. . . . The foreign chemists and manufacturers have looked all around, not merely in their own countries but wherever they could find improved methods and apparatus; and upon the practical knowledge thus gained they have brought to bear the scientific training they had received at their universities and polytechnic schools. Thus they have already in many fields formerly remunerative to British manufacturers distanced the latter, immensely aided though these be by their long occupation of the ground, and by permanent natural advantages, such as cheapness of coal and freight, superior command of capital, etc., and this is likely to go on to an increasing extent if many British chemical manufacturers decline to profit from the scientific study of their respective branches. This is all the less excusable as England from of old has been a stronghold of scientific chemistry, and can hold its own against the whole world in that respect.

Thirteen years later Lunge (1893, p.398) again speculated on the reason for British decline:

> It is notorious that the English alkali trade is no longer rapidly expanding, while that of other industrial nations has made enormous strides. Customs duties and other external circumstances partly account for this, but a very important part is also played by the more careful scientific training of the managers and chemists in the other countries, exceptions on both sides, of course, notwithstanding.

'We are sorry to say', observed the *Chemical Trade Journal* (3 Nov. 1888, p.275), 'that many of the manufacturers of chemical products in this country are absolutely ignorant of what is taking place on other shores, and not only so, but forgetful also of what is happening also in their own commercial centres . . .'. In the industrial regions of Britain there still survived an individualistic, unsophisticated approach to chemicals manufacture, so that it was said of East Lancashire and West Yorkshire in particular that one could set up a works with 'one or two tubs, a few carboys and a stink' (*C.T.J.*, 13 Aug. 1889).

A few years later a survey of technical excellence in chemical manufacture was attempted. The Society of Chemical Industry sent a circular to its members who were managers or chemists in the industry. 502, or half of those addressed, replied. Of these, 107 were graduates, the rest had not taken a degree. Of the 107, 59 were graduates of a British university, 32 of a foreign one, and 16 had studied both at home and overseas. (*C.N.*, 10 Oct. 1902, pp.176–7.) Good management, technical excellence, and good sales organization seem to have been the vital ingredients in Continental success. Levinstein (1886, p.355), the Manchester dyestuffs manufacturer, pointed out that although salt, pyrites, ammonia, lime, coal, and nitrate of soda were all cheaper in Britain, German customers could buy ammonia soda in Frankfurt for £4. 7*s*. 6*d*. a ton, whereas the Manchester customer had to pay £5. 5*s*. Fifteen years later American commentators noted that in the electrolytic alkali process, Germany had no great fuel cost advantage over Britain but had done well by 'attention to details, and constant improvement of processes'.

The Germans themselves puzzled over the reasons for their success. In 1889 Hasenclever had spelt out the difficulties of the German soda industry:

> There is no sound basis in Germany for a permanent export trade in soda. Even a superficial comparison with the position of England, with its well-developed canal system [*sic*] and foreign trade connections, with that of Germany is sufficient to show the difficulties which the latter must overcome in any attempt to compete with England in manufactures requiring the concentration of low value raw materials at one point.

English railway rates were said to be lower owing to the competition of private companies as opposed to the monopolistic position of Prussian state railways. (*C.T.J.*, 26 Oct. 1889, pp.258–9.) By 1902 Hasenclever noted that the only well-located ammonia soda works in Germany was Bernburg, on the Saale, and in the Stassfurt salt basin where brine, lignite, and limestone were on the spot; all other ammonia soda works paid heavy freight on their raw materials. Even so, Germany's ammonia soda production had increased since the mid-1880s much more rapidly even than Brunner Mond's, in spite of the latter's unquestioned excellence (*J.S.C.I.*, 28 Feb. 1902, p.290). Technical progress and the forward impetus of a new, nationally-conscious industrializing power played a large role. In 1894 the German Society for Applied Chemistry stressed the need for both pure chemistry and chemical technology at the universities 'in order that the German chemical industry may in the future occupy that elevated position which forms the envy of other nations at the present day' (quoted *B.P.P.*, 1901, LXXV, p.33). Against that the British had no answer. A few years later, with the advantages of a vast market, no substantial pre-existing alkali industry to care for, and rapid economic growth, the Americans too were rapidly overhauling them.

In 1901 the *Chemical Trade Journal* (9 Mar. 1901, pp.219–20), replying to a correspondent who asked why so many saltcake furnaces were going out of production in Widnes, referred to progress in the United States:

> The American caustic soda works are marvels of engineering skill; no money has been spared to reduce labour to a minimum, and to economize fuel to the utmost farthing. The fault with chemical manufacturing in this country is the difficulty in persuading those interested to spend their money in improved plant. In days gone by, coal was cheap, labour was rough, cheap and ready, and, for the rude methods then involved, any old boilers, second-hand machinery, and cheap supervision was considered quite good enough in the manufacture of chemicals. Our continental neighbours, as well as our American cousins, can now say with pride, 'Nous avons changé tout cela'.

The world dominance of British alkali firms was at an end.

EPILOGUE: OLD ALKALI AND MODERN LANDSCAPES

When building the Black Country link to the M.6 motorway in the late 1960s, the engineering contractors met with 'a giant bluish mound of chemical waste'—in fact the tank waste of Chance's Oldbury works finally closed only in the mid-1960s (*Chemistry and Industry*, 11 Oct. 1969, p.1433). On Tyneside, evidence is more widespread. Fifty years ago it was stated that 'at Bill Quay and elsewhere along the river are to be found the remnants of once-prosperous communities living in dilapidated houses at the foot of enormous slagheaps [*sic*], now covered with coarse grass' (Mess, 1928, pp.39–40). Within a few years all this wasteland had become one of the impediments to the attraction of the new industry that was so desperately needed on Tyneside (Craggs, 1938, xxvii). Much of it still remains, but redevelopment has begun. On the waste and plant site of Allhusens' works there is now an estate for light industry and for extensive warehousing for central Tyneside. Immediately to the east a sports stadium has been established, and the beginnings of a riverside park, on the ground which connected Allhusens' works to the old Friars Goose works. (*Times Trade and Engineering*, Jan. 1936, p.xxxviii; Dec. 1937, p.29.) The old soap firm of Hedleys, now operated by the Proctor and Gamble group, is the chief survivor of the old local outlets for alkali. Names of places and streets bear evidence of Tyneside's once major alkali production: Glasshouse Bridge spanning the Ouseburn, and Bottle House and Glass House Street in St. Peter's, between Byker and Walker. On the Hebburn to Shields Road just opposite the Bede Trading Estate still stands the Alkali Inn in which those who laboured in the awful conditions of the chemical works slaked their thirst. Some three miles to the west in Hebburn is the Argyll Hotel and, further along, St. Rollox Street, names which indicate the considerable importation of Glasgow capital and Scottish labour when Tennants built on Tyneside in the 1860s. Allhusen, Clapham, Losh, Tennant, and Pattinson are all remembered in Tyneside street names—mostly in areas scheduled for redevelopment (Campbell, 1971, p.138). In Scotland the site of St. Rollox works has been built on and its waste heaps covered by great tower blocks in which some of Clydeside's old tenement dwellers have been rehoused.

In Lancastria, the alkali trade left Earlestown (Newton-le-Willows) with a Chemical Street and Vitriol Square. In St. Helens determined efforts have been made since the 1950s to remove chemical waste heaps and to reclaim derelict plant sites. Some of the latter have been used for new industry, one has become a municipal housing estate; and a number of heaps, being rich in lime, have been flattened for playing-fields. Elements of the chemical industry remain, such as Leathers Chemicals. Town employment is still dominated by glass, which was closely associated with the alkali industry from the early days and which in the early 1970s provided some 20,000 out of St. Helen's total of 56,000 jobs.

Widnes was dominated by alkali, and chemicals still play a leading role in its economy. Until the 1950s the Muspratt works, dating from the 1850s, made sodium sulphide in the original black ash revolving furnaces using coal and saltcake. (Hardie, 1950, p.229.) Product lines of old works evolved. The Pilkington-Sullivan works, established by Pilkingtons of St. Helens in 1865 to provide alkali for their glass-works, had become by the late 1960s the paraquat-producing unit of the I.C.I. Mond Division. The research laboratory built by Hurter in the early 1890s survives as a research establishment of the Mond Division. Within the town, remnants of the old alkali trade are everywhere. In the early 1960s, before the engineering works for the new Runcorn Bridge were begun, frightful waste heaps hemmed it in. The bridge across the Sankey Canal gave an excellent view of waste grounds and black factory silhouettes. Along the Ditton Road, firms, which have now disappeared, had colonized the old alkali sites. In the older parts of the town, the streets of low dark housing are reminiscent of central Middlesbrough, though on a much smaller scale. Street names recall alkali pioneers—Deacon Road, Gossage Street.

Within the Cheshire salt-field, active chemical works triumphantly carry on the alkali technology of the late nineteenth century at a vastly extended scale into the late twentieth century. Few industrial prospects in Britain can be more impressive or forbidding than that of chemical Northwich viewed from the main road near the Anderton Lift. In Winnington, the pioneer Solvay town, Brunner Mond & Co. followed a modified, more enlightened, but essentially similar development process to that of their Leblanc predecessors and rivals. In 1882, extending their workers' cottage developments they produced Solvay Road, Hemming Street, and Dyar Terrace. During the twenty-fifth anniversary celebrations of the firm it was recorded that Mond walked over the Weaver and the canal bridges and up Anderton Hill. Looking back for a long time over to Winnington Park and to his works there, he remarked 'And this is the great Babylon which I have builded' (Tangye, 1948). Eighty years later the same view shows how firmly the chemical foundations were laid.

REFERENCES

AFFLECK, J. (1877). Presidential Address. *T.N.C.S.* 4, 1877–80.
Alkali Inspectors. Annual Report, *B.P.P.*
Alkali Statistics (1882). *B.P.P.*
ALLEN, J. F. (1881). *J.S.C.I.*, 29 June.
— — (1907). *Some Founders of the Chemical Industry.*
ALLHUSEN, C. (1886). *Evidence to Royal Commission on the Depression of Trade and Industry.*
ARMSTRONG, W. G. (ed.) (1864). *Industrial Resources of the Tyne, Wear and Tees.*
ARROWSMITH (1815). *Map of England and Wales.*
ASHMORE, O. (1969). *Industrial Archaeology of Lancashire.*
AYRSHIRE (1951). *Third Statistical Account of Scotland.*

BAILEY, J. (1810). *General View of the Agriculture of County Durham.*
BAINES, E. (1825). *History, Directory and Gazetteer of the County Palatine of Lancashire,* vol. I.
BANNERMAN, H. (1897). *Industrial Lancashire.*
BARKER, T. C. (1960). *Pilkington Brothers and the Glass Industry.*
—— and HARRIS, J. R. (1954). *A Merseyside Town in the Industrial Revolution. St. Helens 1750 to 1900.*
—— DICKINSON, R., and HARDIE, D. W. F. (1972). 'The Origins of the Synthetic Alkali Industry in Britain' in S. Lierman, *Europe and the Industrial Revolution* (1972).
BELL. Minute Books of Bell Brothers.
BELL, I. L. (1862). Evidence to Select Committee on Noxious Vapours, *B.P.P.*
—— (1868). *T.N.C.S.* I.
—— (1869). *T.N.C.S.* I.
—— (1871). in *Report of the Coal Inquiry*, II.
—— (1873). Presidential Address to the Iron and Steel Institute, *J.I.S.I.*, 1873.1
—— (1881). *Procs. Newcastle Meeting of Inst. of Mechanical Engineers.*
—— (1887). 'On the Manufacture of Salt near Middlesbrough', *Procs. Inst. of Civil Engineers,* XC, 1886/7, Part 4
—— and BURNETT, J. C. (1885). *Report of the Proceedings and Papers of the Industrial Remuneration Conference*, January.
BELL, I. T. (1849). Plan of the River Tyne.
BEVAN, G. P. (1876). *The Industrial Classes and Industrial Statistics. Mining, Metals, Chemicals, Ceramics, Glass and Paper.*
BLATCHFORD, R. (1899). *Dismal England.*
BLOUNT, B. (1896). 'On the manufacture of alkali and bleach by chemical and electrolytic methods', *Eng* 18 and 25 Dec.
Board of Trade (1845). 'Report of the Railway Department of the Board of Trade on schemes for extending railway communication in Lancashire and adjoining districts', *B.P.P.* XXXIX.
BOURNE, T. and HORNE, W. (1817). Evidence to Select Committee on Rock Salt in the Fisheries, *B.P.P.* III, 247, pp.22, 24.
British Association (1838). Report of Newcastle upon Tyne Meeting.

—— (1863). *Trans. of sections, Newcastle-upon-Tyne meeting.*
—— (1889). *Handbook to the Industries of Newcastle and District.*
—— (1896). *Handbook to Liverpool.*
—— (1923). *Excursions and Works visits, Liverpool Meeting,* Visit to Pilkington-Sullivan Works.
BROCK, J. and DENNIS, J. H. (1890). Circular of Liverpool alkali manufacturers to firms in the trade. I.C.I., London Records.
BRUNNER, J. (1872). Cost estimates, quoted Watts (1923), p.102.
BRUNNER, Sir J. (1896). Address to the half-yearly ordinary meeting of Brunner, Mond and Co., *C.T.J.*, 27 Feb.
Brunner Mond Record (periodical).
BUTT, J. (1967). *The Industrial Archaeology of Scotland.*

CALVERT, A. F. (1915). *Salt in Cheshire.*
CALVERT, R. C. (1871). 'An Account of the Manufacture of Copper on the Tyne', *CN*, 20 Nov.
CAMERON, J. (1845). *New Statistical Account XIV. Inverness and Ross and Cromarty.*
CAMPBELL, W. A. (1962). *The Old Tyneside Chemical Trade.*
—— (1971). *The Chemical Industry.*
CASSELL (1882). *Great Industries of Great Ritain.* 3 vols.
CASTNER-KELLNER (1945). *Fifty Years of Progress. The Story of the Castner-Kellner Company.*
CHAMBERS (1881). *Encyclopaedia.*
CHANCE, H. (1856). 'On the manufacture of Crown and Sheet Glass', *J.S.A.* 4.
CHANCE and HUNT (1875). Process sheets. I.C.I., Runcorn Records, C.H. 21.
—— (1892). Agreement with United Alkali Company for 1892. I.C.I., Runcorn Records.
Chance and Hunt, Carriage (n.d.). Rates of Carriage Book. I.C.I., Runcorn Records.
Chance and Hunt, History (n.d.). A Short History. I.C.I., London Records, Box 12. 1. 8.A.
Chance and Hunt, Statistics (n.d.). I.C.I., Runcorn Records, C.H./54.
CLAPHAM, R. C. (1868). *Evidence to Select Committee on Scientific Instruction.*
—— (1869). 'An Account of the Soda Manufacture on the Tyne', *T.N.C.S.* I.
—— (1874). 'Notes on a visit to Chemical Works in France', *C.N.*, 10 Dec.
—— (1878). Presidential Address, *T.N.C.S.* 4.
CLAPHAM, Sir J. (1938). *The Economic History of Modern Britain,* III: *Machines and National Rivalries.*
CLARENCE (1896). Clarence Soda ash cost statement, Brunner Mond Sundry Costs (49). I.C.I., Runcorn Records.
CLOW, A. and N. (1947). 'The Natural and Economic History of Kelp', *Annals of Science*, 5, No. 4, 15 July.
Coal (1871). Report of Commission of Inquiry.
Coal (1873). Select Committee, Minutes of Evidence.
COHEN, J. (1956). *The Life of Ludwig Mond.*
Commissioners of Inquiry (1845). *Second Report of Commissioners of Inquiry into the state of large Towns and Populous Districts.*
Comprehensive Gazetteer (1895). *of England and Wales.*
CONSTABLE, A. (1822). 'Newcastle-upon-Tyne' in *The Edinburgh Gazetteer or Geographical Dictionary.*
CRAGGS, J. W. (1938). 'Chemical Industries on the North East Coast', *Times Trade and Engineering*, October.

DAUMAS, M. (1969). *A History of Technology and Invention: Progress through the Ages,* vol. 2.
DAVIDSON, J. E. (1923). 'The Tyne Alkali Industry' in U.A.C. (1923).

DAVIDSON, R. H. (1932). 'History of United Alkali Company' (unpublished).
DAVIES, M. L. (1894). Report on the Liverpool Works, 16 February. I.C.I., London Records, Box 12. 1. 86.
DAVIS (1913). *The Chemical Trade Directory*, 5th edition.
Depression of Trade (1886). Royal Commission on Depression of Trade and Industry.
DICK, W. F. L. (1973). *A Hundred Years of Alkali in Cheshire*.
DITTMAR, W. and PATON, J. (1887a). 'Soap' in *Encyclopaedia Britannica*, 9th edition.
—— (1887b). 'Sodium' in *Encyclopaedia Britannica*, 9th edition.
Dresden (1885). Report by H.M. Chargé d'Affaires at Dresden on the effects of the German Customs Tariff Reform of 1879 and on the revision of 1885, *B.P.P.*, 1884/5, vol. 37.
Dumbarton (1845). *New Statistical Account*.

EDMONDSTON, L. (1845). 'General Observations' in *New Statistical Account*, Shetland.
Electrolysis (1893/4). Correspondence between United Alkali Company and the Aluminium Company on electrolytic processes. I.C.I., London Records.

FARDON, J. (1845). Evidence to the Select Committee on the Oxford, Worcester and Wolverhampton Railway Bill, *B.P.P.* Committee Report, 5.
FERGUSON, J. (1876). Chemical Manufactures, in *Notes on some of the principal manufactures of the west of Scotland*.
FINDLAY, A. (1965). *A Hundred Years of Chemistry*.
FISHER UNWIN, T. (1888). *Industrial Rivers of the United Kingdom*.
Fleetwood (1903). Fleetwood Manager to J. A. Rayner. I.C.I., London Records, Box 12.1 8B.
—— (1963). 'Reminiscences of the Chemical Industry in Fleetwood and Widnes, 1901–1928'. I.C.I., Runcorn Records, M.D. 2036.
FLETCHER, A. E. (1887). 'The Present Position of the Alkali Trade', *Industries*, 16 Sept. *Remarks of F. Hurter*.
—— (1888). 'The Present State of the Law concerning the pollution of air and water', *C.T.J.*, 28 Apr.
FORDYCE, W. (1857). *The History and Antiquities of the County Palatine of Durham*.
—— (n.d.). *History of Tynemouth*.
FORSTER, G. B. and BELL, M. W. (1870). *I.C.T.R.*, 16 Mar.
FULLARTON, A. (1842). *Parliamentary Gazetteer*.
—— (1859). *A Gazetteer of the World or Dictionary of Geographical Knowledge*.

GAMBLE, D. (1878). Select Committee on Noxious Vapours, *B.P.P.*
General Chemicals (1963). Notes on the Ammonia Soda Co. Ltd. I.C.I., Runcorn Records.
GIBSON, A. M. (1893). 'The History of Alkali Manufacture in Great Britain' in Rothwell (1893).
GIBSON, W. S. (1860). *Memoir on Northumberland*.
GILLESPIE, C. C. (1957). 'The Discovery of the Leblanc Process', *Isis* 48.
GOODRICH-FREER, A. (1902). *Outer Isles*.
GOODWIN, J. (1842). 'Progress on the Two Sicilies under the Spanish Bourbons from the Year 1734/5 to 1840', *J.S.S.* 5.
GOSSAGE, W. (1861). 'On the History of the Alkali Manufacture', *Trans. of sections, British Association, Liverpool*.
—— (1862). Evidence to Select Committee on Noxious Vapours, *B.P.P.*
Great Exhibition (1852). Report of Jurors and catalogue.
GRINDON, L. (1892). *Lancashire, Brief Historical and Descriptive Notes* (mostly written 1881).

HABER, L. F. (1958). *The Chemical Industry during the Nineteenth Century*.

HÄGERSTRAND, T. (1973). 'The Domain of Human Geography' in R. Chorley, *Directions in Geography*.

HARDIE, D. W. F. (1950). *A History of the Chemical Industry in Widnes*.

—— (1961). *The Chemical Industry of Merseyside*.

—— and PRATT, J. D. (1966). *A History of the Modern British Chemical Industry*.

HAWES, W. (1856). 'On the Manufacture of Soap', *J.S.A.* IV.

HILL, M. F. (1964). *Magadi: The Story of the Magadi Soda Co.*

HOHENBERG, P. W. (1967). *Chemicals in Western Europe 1850–1914*.

HOU, T. P. (1942). *The Manufacture of Soda, with special reference to the Ammonia Soda Process*.

HOWARD, H. (1866). 'On the process and manufacture of plate glass in England, France, Belgium and Russia', *J.S.S.* 29.

HURTER, F. (1888). *The Philosopher's Stone*.

—— (1898). A Memorandum on Electrolysis prepared for the private use of the Directors of the United Alkali Company Ltd.

HUSKISSON, H. C. (1853). 'Salt—the Sources from whence it is obtained and the processes involved in its manufacture', *J.S.A.* 1, 1852–3.

HYDE, F. E. (1971). *Liverpool and the Mersey. An Economic History of a Port, 1700 to 1970*.

Imperial (1872). *Gazetteer of England and Wales*.

Imperial Gazetteer (1874).

I.C.I. (1948). *Alkali News*, 75th Anniversary Issue.

—— (1951). *Winnington Hall*.

I.C.I., Stoke. Runcorn Records.

I.C.I., Netham. A Short History of Netham Works. I.C.I., London Records.

I.C.I., Salt. Records relating to the South Durham salt field in the 1920s. I.C.I., Runcorn Records.

IRVINE, A. S. (1958). *A History of the Alkali Division of I.C.I.*

JEANS, J. S. (1877). *Notes on Northern Industries*.

KERSHAW, J. B. C. (1897). Quoted *C.T.J.*, 1 May.

—— (1906). 'The Future of Leblanc Alkali Works', *C.T.J.*, 31 Mar.

KINGZETT, C. T. (1877). *The History, Products and Processes of the Alkali Trade*.

KNIGHT, C. (1851). *Knight's Encyclopaedia of the Industry of All Nations*.

—— (1855). *English Cyclopaedia: Geography*.

—— (1860). *English Cycopaedia: Art and Science*, 3 and 4.

—— (1873). *The English Cyclopaedia: Supplement*.

Labour (1893/4). Report on the Conditions of Labour in Chemical Works, *B.P.P.*

Lanark (1845). *New Statistical Account*.

LANDES, D. (1965). *Technological Change and Development in Western Europe 1750–1914*.

LAWTON, R. (1953). in Smith (1953).

LEIFCHILD, J. R. (1843). Report on glasshouses, potteries, paper mills and miscellaneous manufactures in Northumberland, and Durham north of the river Wear, *Children's Employment Commission 2nd Report*, Appendix.

LEVI, L. (1865). 'On the Economic Conditions of the Highlands and Islands of Scotland', *J.S.S.* 28.

LEVINSTEIN, I. (1886). 'Observations and suggestions on the present position of the British chemical industries with special reference to coal tar derivatives', *J.S.C.I.*, 29 June.

LEWIS (1831, 1833, 1848). *Topographical Dictionary of England*.

—— (1847). *Topographical Dictionary of Scotland*.

LEWIS, D. L. (1911a). *Widnes—A Review and a forecast of a municipality in the making*, quoted J. Morgan (1972).
—— (1911b). Address to the North West Sanitary Inspectors Association, Widnes meeting, quoted *Widnes Weekly News*, 17 Mar.
LITTLE, E. S. (1915/16). Review of Trade, Brunner, Mond & Co., China. I.C.I., London Records.
LOMAS, J. (1880). *A Manual of the Alkali Trade*.
LUNGE, G. (1880). *Sulphuric Acid and Alkali*, vol. 3.
—— (1893). 'The Education of Industrial Chemists—a paper to the Chemists' Congress, Chicago', *C.T.J.*, 16 Dec.
—— (1910). 'Alkali Manufacture' in *Encyclopaedia Britannica*, 11th edition.

McCULLOCH, J. R. (1839). *A Statistical Account of the British Empire*, 2nd edition.
—— (1841). *A Dictionary, Geographical, Statistical and Historical*, article 'Hebrides'.
McGOWAN, Lord (1948). Speech on the 75th Anniversary of the foundation of Winnington works.
MACKENZIE, E. and ROSS, M. (1834). *An Historical, Topographical and Descriptive View of the County Palatine of Durham*, vol. I.
MACROSTY, H. W. (1907). *The Trust Movement in British Industry*.
MACTEAR, J. (1868). Lecture to Philosophical Society of Glasgow, *C.N.*, 8 May.
—— (1877). 'On the growth of the alkali and bleaching powder manufacture of the Glasgow district', *C.N.*, 5 Jan.
—— (1881). 'On Mechanical furnaces employed in the alkali manufacture'. First General Meeting, Society of Chemical Industry.
MANNEX, P. (1855). *History, Topography and Directory of Mid-Lancashire*.
MARTIN, G., SMITH, S., and MILSON, F. (1916). *The Salt and Alkali Industry*, Manuals of Chemical Technology, VI.
MASON, W. (1908). 'The Manufacture of Ammonia Soda', *C.T.J.*, 7 Nov.
MESS, H. A. (1928). *Industrial Tyneside: A Social Survey*.
MIDDLEBROOK, S. (1950). *Newcastle-upon-Tyne: Its Growth and Achievement*.
MILLWARD, R. (1955). *Lancashire*.
Mineral Statistics of the United Kingdom (annual).
MINSTREL (1906). 'A Widnes Garland of Verses', reprinted from the *Widnes Weekly News*.
MOND, A. (1903). 'The Alkali Industry', in H. Cox, *British Industries under Free Trade*.
MOND, L. (1869). Evidence to Rivers Pollution Committee, *B.P.P.*, 1870, XL.
—— (1872 and 1888). Estimates of Production costs in Brunner, Mond/49, Sundry Costs Ledger. I.C.I., Runcorn Records.
—— (1872b). Cost Estimates, quoted Cohen (1956).
—— (1875). Letter. I.C.I., Runcorn Records.
—— (1884). Costs estimates, Brunner Mond, Sundry Costs Ledger, I.C.I., Runcorn Records.
—— (1896). 'The History of the Manufacture of Chlorine', *Eng.*, 6 Nov.
MORGAN, J. (1972). *'A Study of the Population of Widnes in the Nineteenth Century'*, un published B.Ed. thesis, Padgate College of Education.
MORGAN, K. (n.d.). On Bristol works. I.C.I., London Records, Box 12. 1. 8b.
MORLEY, J. (1863). *Transactions, North of England Institute of Mining Engineers*, 13.
MORRISON, J. (1874). 'On the Manufacture of Caustic Soda', *T.N.C.S.* III.
—— (1875a). 'On the Manufacture of Caustic Soda', *C.N.*, 26 Feb.
—— (1875b). *C.N.*, 19 Mar.
MUMFORD, L. (1934). *Technics and Civilisation*.
MUSPRATT, E. (1881). Evidence to Select Committee on Railways, *B.P.P.* XIII.
—— (1886). 'On the Development of thealkali and Associated Industries', *J.S.C.I.*, 29 July.
MUSSON, A. E. (1965). *Enterprise in Soap and Chemicals: Joseph Crosfield & Sons Ltd. 1815–1965*.

National (1847/8). *The National Cyclopaedia.*

NICKSON, C. (1887). *History of Runcorn.*

NORMAN, Sir F. (1931). 'A Short History of the part Cheshire Works have played in the development of the Chemical Industry'. I.C.I., London Records.

Noxious Vapours (1862). Report and Minutes of Evidence of the Select Committee of the House of Lords on injury from . . .

Noxious Vapours (1878). Report of Royal Commission.

OLIVER, T. (1831). *A Picture of Newcastle-upon-Tyne.*

O.S. (1842). First Edition of the Ordnance Survey Map of South Lancashire.

O.S. (1875). Ordnance Survey. First Six Inch Survey of County Durham.

PARTINGTON, J. R. (1919). *The Alkali Industry.*

—— (1964). *A History of Chemistry*, vol. 4.

PATTINSON, H. L. (1853). Address to Felling Chemical Works Reading Society.

PEACOCK, W. H. (1869). 'On the New Red Sandstone of Cleveland and the Rock Salt discovered in it', Paper to Cleveland Literary and Philosophical Society, reported *Eng.*, 19 Nov. 1869.

PENNOCK, J. D. (1903). 'Progress of the Soda Industry in the U.S.A. since 1900', V. *Internationaler Kongress für Angewandt Chemie*, Berlin, Berichte I.

Penny Magazine (1844). 'A Day at the Felling Chemical Works, Newcastle, 25th May.' 'A Day at the Tyne factories, 31 August.'

PILBIN, P. (1935). 'Geographical Studies of the Development of certain industries of Tyneside and North East England', unpublished M.Sc., University of Durham.

PLUMMER, B. (1874). *Newcastle upon Tyne: Trade and Manufactures*

POLACK, W. G. (n.d.). Early History of Alkali in Bristol. I.C.I., London Records, Box 12.1.8b.

Pollution Commission (1872). Rivers Pollution Commission, *B.P.P.* XL.

Popular (1892). *Encyclopaedia.*

PORTER, G. R. (1836). *The Progress of the Nation*, Section V.

Post Office (1863/4). *Directory of Glasgow.*

PROCTOR, B. S. (1883). Inaugural Address to the Newcastle Section of the Society of Chemical Industry, *J.S.C.I.* 2.

Railways (1881). Report of Select Committee on, *B.P.P.* XIII.

RAWSON, R. W. (1839/40). 'On the Sulphur Trade of Sicily and the commercial relations between that country and Great Britain', *J.S.S.* 2.

READER, W. J. (1970). *Imperial Chemical Industries. A History*, I: *The Forerunners, 1870–1926.*

REILLY, D. (1951). 'Salts, Acids and Alkalis in the Nineteenth Century. A Comparison between Advances made in France, England and Germany', *Isis*, 42.

RHODES, E. (1909). Estimates of Alkali Waste around Widnes, quoted Sherlock (1922), p.209.

RICHARDS, P. S. (1973). 'Deeside Industries', unpublished M.Phil., University of London.

RICHARDSON, T. and WATTS, H. (1863). *Chemical Technology, Acids, Alkalies and Salt.*

ROSCOE, H. (1868). Evidence to Select Committee on Scientific Instruction, *B.P.P.*

—— (1881). *Procs. First General Meeting of the Society of Chemical Industry.*

ROSE, F. (1901). Reports on Chemical Instruction in Germany and the Growth and Present Condition of the German Chemical Industries, *B.P.P.* LXXV.

ROTHWELL, R. P. (1893). *The Mineral Industry: Its Statistics, Technology and Trade*, vol. 2.

—— (1895). *The Mineral Industry: Its Statistics, Technology and Trade*, vol. 3.

SAINT FOND, B. F. de (1907). *A Journey through England and Scotland to the Hebrides* (journey made 1784).

St. Helen's Lantern (1887). 'Reflections on the Labours, Recreations and Distractions of her people', 7 Apr.

St. Rollox (1893). Costs of Acid Making at Scottish Works. I.C.I., Runcorn Records.

Salt Duties (1818). *Report from Select Committee on the Laws relating to the Salt Duties.*

Sandbach (1878). Letter from W. A. Richards to Brunner Mond. I.C.I., London Records, 12.1.6A.

SCHUNK, ROSCOE and SMITH (1861). 'On the Chemical Manufactures of South Lancashire', *C.N.*, 21 Sept.

SCRIVENOR, H. (1849). *The Railways of the United Kingdom.*

SHERARD, R. H. (1897). *The White Slaves of England*: being a true picture of certain social conditions in the kingdom of England in the year 1897.

SHERLOCK, R. L. (1921). *Rock Salt and Brine.*

—— (1922). *Man as a Geological Agent.*

SMITH, A. (1867). Report of the Alkali Inspector, quoted *C.G.*, 8 Aug. 1868, p.114.

SMITH, W. (ed.) (1953). *A Scientific Survey of Merseyside.*

SOLVAY, E. (1903). 'A Retrospective Glance at the Process for manufacturing Ammonia Soda', quoted Watts (1923).

SPICER, A. (1903). On paper in H. Cox, *British Industries under Free Trade.*

SPONS (1879). *Encyclopaedia of the Industrial Arts, Manufacture and Commercial Products*, vol. I.

STANFORD, E. C. C. (1884). *J.S.C.I.*, 29 May.

STEVENSON, J. C., CLAPHAM, R. C. and RICHARDSON, T. (1863). 'Chemical Manufactures on the Tyne', *C.N.*, 12 Sept.

STUART, T. W. (1888). 'On the Tees Salt Field', *J.S.C.I.*, 31 Oct.

—— (1889). 'Chemical Manufactures' in British Association, *Handbook to the Industries of Newcastle and District.*

—— (1911). Report on the Tank Waste Deposits at St. Rollox Works to the Health Committee of Glasgow Corporation.

—— (1932). List of Works on the Tyne in 1866. I.C.I., Runcorn Records.

Sulphuric Acid (1918). *Report of a Departmental Committee on the Post-war Position of the Sulphuric Acid and Fertiliser Trades*, Cd 8994.

Sunderland (1886). Submissions of Chambers of Commerce to the Royal Commission on Depression of Trade and Industry, I.

SWAN, J. W. (1898). Inaugural Address, Institute of Electrical Engineers, *Eng.*, 28 Jan.

TANGYE, A. W. (1948). Main Occurrences and Personal Experiences at Brunner Mond & Co. 1885–1926. I.C.I., London Records, Box 12.4.9.A.

TENNANT, E. W. D. (1922). *A Short Account of the Tennant Companies 1797–1922.*

TENNANT, J. (1897). Obituary Notice in *Procs. Inst. Mechanical Engineers.*

Third Statistical Account, Glasgow (1958).

THORPE, T. E. (1895). Presidential Address, *J.S.C.I.*, 28 Feb.

U.A.C. (1890). *Prospectus of United Alkali Company.*

U.A.C. (1896). Report on Balances and Costs of Manufacture. I.C.I., London Records.

U.A.C. (1896b). Widnes Works Production 1891 and 1895. I.C.I., Runcorn Records.

U.A.C. (1900A). Report of London Discussions with Brunner Mond & Co. and Solvay & Co. I.C.I., London Records.

U.A.C. (1900B). Agreement of 30 July between U.A.C., Brunner Mond & Co. and Solvay & Co. I.C.I., London Records.

U.A.C. (1910). Letter to F. H. Hornblower acting for the Birmingham Corrugated Iron Company. I.C.I., Runcorn Records.

U.A.C. (1911). Statistics of Works Capacities. I.C.I., Runcorn Records.
U.A.C. (1919). Agreement as to certain products and markets of Brunner Mond, Castner-Kellner Company and U.A.C., I.C.I., London Records.
U.A.C. (1923). *Centenary of the Alkali Industry 1823–1923.*
U.A.C. (1925/1926). Comparisons of Works. I.C.I., London Records.
U.A.C. (1930/1932). List of Works of the U.A.C.
U.A.C., Liverpool. Minute Books of U.A.C. Directors Meetings. I.C.I., Runcorn Records.
U.A.C., Tyne. Minutes of the Tyne Branch of U.A.C., I.C.I., London Records.
URE, A. (1853). *Dictionary of Arts, Manufactures and Mines.*

VINCENT, C. W. (1874). 'On some recent processes for the manufacture of soda', *J.S.A.*, 17 Apr.
—— (*c.*1877). (Editor), *Chemistry Theoretical, Practical and Analytical as Applied to the Arts and Manufactures.*

WALKER, J. W. (n.d.). Letter to J. Brock of U.A.C. on Black Ash costs. I.C.I., Runcorn Records.
WALLWORK, K. L. (1967). 'Map Interpretation and Industrial Location: The Example of Alkali Manufacture in Lancashire', *Geography*, 62, 2 Apr.
WARD, G. (1883). 'A Short Notice of the Salt Deposits near Middlesbrough', *J.S.C.I.*, 29 June.
WATTS, J. I. (1918). 'On Ludwig Mond', *Trans. Chemical Society*, 113.
—— (1923). *Brunner, Mond, The 50th Anniversary 1873–1923.*
WELDON, W. (1874). 'On the Manufacture of Chlorine', *J.S.A.*, 29 May.
—— (1883). *J.S.C.I.*, 8 Jan.
—— (1884). to Newcastle-upon-Tyne section of the Society of Chemical Industry, *M.J.*, 19 July.
WELTON, T. A. (1897/8). 'On Forty Years' Industrial Change in England and Wales', *Trans. Manchester Statistical Society.*
WHELLAN, W. (1856). *History, Topography & Directory of the County Palatine of Durham.*
Whitaker (1913). *Almanack.*
WHITE, W. (1859). *Northumberland and The Border.*
Widnes (1872). Minutes of Evidence of the House of Commons Committee on the Widnes Railway Bill. *B.P.P.*
Widnes (1899). Annual Report of the Medical Officer of Health.
Widnes Guide (1910).
Widnes Foundry (n.d.). *A History of the Widnes Foundry and Engineering Co. Ltd.*
WILSON, J. M. (1872). 'Liverpool' in Imperial Gazetteer of England and Wales.
WITT, O. N. (1909). 'Evolution in Applied Chemistry', *Procs. 7th International Congress of Applied Chemistry*, London.
WORRALLS (1876). *Directory of Warrington, Wigan, St. Helens and District.*
WORSLEY, J. (1876). Letter to Alexander M. Chance. Chance and Hunt Caustic Soda Folder, I.C.I., Runcorn Records.
WYATT, F. (1893). 'The Progress of the Chemical Industry' in Rothwell (1893).

Index